Methods in Cell Biology
The Immunological Synapse Part A

Volume 173

Series Editor

Lorenzo Galluzzi
*Weill Cornell Medical College,
New York, NY, United States*

Methods in Cell Biology

The Immunological Synapse Part A

Volume 173

Edited by

Clément Thomas
Luxembourg Institute of Health, Luxembourg City, Luxembourg

Lorenzo Galluzzi
Weill Cornell Medicine, New York, NY, United States

Academic Press is an imprint of Elsevier
50 Hampshire Street, 5th Floor, Cambridge, MA 02139, United States
525 B Street, Suite 1650, San Diego, CA 92101, United States
The Boulevard, Langford Lane, Kidlington, Oxford OX5 1GB, United Kingdom
125 London Wall, London, EC2Y 5AS, United Kingdom

First edition 2023

Copyright © 2023 Elsevier Inc. All rights reserved.

No part of this publication may be reproduced or transmitted in any form or by any means, electronic or mechanical, including photocopying, recording, or any information storage and retrieval system, without permission in writing from the publisher. Details on how to seek permission, further information about the Publisher's permissions policies and our arrangements with organizations such as the Copyright Clearance Center and the Copyright Licensing Agency, can be found at our website: www.elsevier.com/permissions.

This book and the individual contributions contained in it are protected under copyright by the Publisher (other than as may be noted herein).

Notices
Knowledge and best practice in this field are constantly changing. As new research and experience broaden our understanding, changes in research methods, professional practices, or medical treatment may become necessary.

Practitioners and researchers must always rely on their own experience and knowledge in evaluating and using any information, methods, compounds, or experiments described herein. In using such information or methods they should be mindful of their own safety and the safety of others, including parties for whom they have a professional responsibility.

To the fullest extent of the law, neither the Publisher nor the authors, contributors, or editors, assume any liability for any injury and/or damage to persons or property as a matter of products liability, negligence or otherwise, or from any use or operation of any methods, products, instructions, or ideas contained in the material herein.

ISBN: 978-0-323-90154-3
ISSN: 0091-679X

For information on all Academic Press publications
visit our website at https://www.elsevier.com/books-and-journals

Publisher: Zoe Kruze
Developmental Editor: Naiza Ermin Mendoza
Production Project Manager: Vijayaraj Purushothaman
Cover Designer: Matthew Limbert

Typeset by STRAIVE, India

Contents

Contributors xi

CHAPTER 1 Imaging polarized granule release at the cytotoxic T cell immunological synapse using TIRF microscopy: Control by polarity regulators 1
Marie Juzans, Céline Cuche, Vincenzo Di Bartolo, and Andrés Alcover

    1. Introduction 2
    2. Materials 4
    3. Methods 5
    4. Concluding remarks 11
    5. Notes 11
    Acknowledgments 12
    Declaration of interests 12
    References 12

CHAPTER 2 Analysis of centrosomal area actin reorganization and centrosome polarization upon lymphocyte activation at the immunological synapse 15
Sara Fernández-Hermira, Irene Sanz-Fernández, Marta Botas, Victor Calvo, and Manuel Izquierdo

    1. Introduction 17
    2. Materials, cells, immunological synapse formation and image capture 20
    3. Imaging the immunological synapse 21
    4. Discussion, future perspectives and concluding remarks 25
    5. Notes 27
    Acknowledgments 29
    Conflict of interest 29
    Author contributions 29
    References 29

CHAPTER 3 P815-based redirected degranulation assay to study human NK cell effector functions 33
Iñigo Terrén, Gabirel Astarloa-Pando, Ainhoa Amarilla-Irusta, and Francisco Borrego

1. Introduction 34
2. Prior to the assay 37
3. Redirected degranulation assay 37
4. Flow cytometry staining 38
5. Analyzing CD16-induced effector functions of NK cells 41
6. Concluding remarks 44
7. Notes 44
Acknowledgments 46
Conflict of interests 46
References 46

CHAPTER 4 Cytotoxic and chemotactic dynamics of NK cells quantified by live-cell imaging 49
Yanting Zhu and Jue Shi

1. Introduction 50
2. Materials 51
3. Methods 53
4. Notes 62
Acknowledgments 63
References 63

CHAPTER 5 Quantification of interaction frequency between antigen-presenting cells and T cells by conjugation assay 65
Ondrej Cerny

1. Introduction 66
2. Materials 67
3. Methods 68
4. Concluding remarks 72
5. Notes 72
Acknowledgment 74
References 74

CHAPTER 6 Assessment of membrane lipid state at the natural killer cell immunological synapse 77
Yu Li and Jordan S. Orange

1. Introduction 78
2. Materials 80
3. Common procedures 81
4. Assessment of membrane lipid state via confocal microscopy ... 82
5. Assessment of membrane lipid state via total internal reflection fluorescence microscopy 84
6. Data analysis 86
7. Notes 87
Acknowledgments 88
References 88

CHAPTER 7 Study of the effects of NK-tumor cell interaction by proteomic analysis and imaging 91
Chiara Lavarello, Paola Orecchia, Andrea Petretto, Massimo Vitale, Claudia Cantoni, and Monica Parodi

1. Introduction 93
2. Melanoma/NK cell co-culture 93
3. Cell imaging 97
4. Analysis of the proteomic changes related to the process of EMT induced by NK cells 99
Notes 105
Acknowledgments 106
Declaration of interests 107
References 107

CHAPTER 8 Protocol for the murine antibody-dependent cellular phagocytosis assay 109
Eliana Stanganello, Magdalena Brkic, Steven Zenner, Ines Beulshausen, Ute Schmitt, and Fulvia Vascotto

1. Introduction 110
2. Materials 112
3. Notes 118
Acknowledgment 118
Author disclosure 118
References 118

CHAPTER 9 Quantification of lymphocytic choriomeningitis virus specific T cells and LCMV viral titers 121
Melanie Grusdat, Catherine Dostert, and Dirk Brenner

1. Introduction 122
2. Materials 123
3. Methods 124
4. Notes 129
5. Conclusion 129
Acknowledgments 130
References 130

CHAPTER 10 An in vitro model to monitor natural killer cell effector functions against breast cancer cells derived from human tumor tissue 133
Nicky A. Beelen, Femke A.I. Ehlers, Loes F.S. Kooreman, Gerard M.J. Bos, and Lotte Wieten

1. Introduction 135
2. Before you begin 137
3. Key resources table 138
4. Materials and equipment 139
5. Step-by-step method details 140
6. Expected outcomes 148
7. Quantification and statistical analysis 148
8. Advantages 151
9. Limitations 151
10. Safety considerations and standards 151
References 152

CHAPTER 11 Standardized protocol for the evaluation of chimeric antigen receptor (CAR)-modified cell immunological synapse quality using the glass-supported planar lipid bilayer 155
Jong Hyun Cho, Wei-chung Tsao, Alireza Naghizadeh, and Dongfang Liu

1. Introduction 156
2. Before you begin 158
3. Key resources table 159
4. Materials and equipment 160

5. Step-by-step method details 161
6. Expected outcomes 167
7. Statistical analysis 169
8. Limitations 169
References 170

CHAPTER 12 Potency monitoring of CAR T cells 173
Dongrui Wang, Xin Yang, Agata Xella, Lawrence A. Stern, and Christine E. Brown

1. Introduction 174
2. Preparation of cells 175
3. Setup co-culture for extended long-term killing (ELTK) assay (Assay #1) 176
4. Setup co-culture for *re*-challenge assay (Assay #2) 178
5. General procedures of flow cytometry analysis 179
6. Analysis of CAR T cell killing and T cell counts by flow cytometry 179
7. Analysis of CAR T cell phenotypes by flow cytometry 181
8. Quantification of cytokine production by CAR T cells 185
9. Concluding remarks 185
10. Notes 186
References 187

Contributors

Andrés Alcover
Institut Pasteur, Université Paris Cité, INSERM-U1224, Unité Biologie Cellulaire des Lymphocytes, Ligue Nationale Contre le Cancer, Équipe Labellisée Ligue-2018, Paris, France

Ainhoa Amarilla-Irusta
Biocruces Bizkaia Health Research Institute, Immunopathology Group, Barakaldo, Spain

Gabirel Astarloa-Pando
Biocruces Bizkaia Health Research Institute, Immunopathology Group, Barakaldo, Spain

Nicky A. Beelen
Department of Internal Medicine, Division of Hematology, Maastricht University Medical Center; GROW-School for Oncology and Reproduction, Maastricht University; Department of Transplantation Immunology, Tissue Typing Laboratory, Maastricht University Medical Center, Maastricht, The Netherlands

Ines Beulshausen
TRON - Translational Oncology at the University Medical Center of the Johannes Gutenberg University GmbH, Mainz, Germany

Francisco Borrego
Biocruces Bizkaia Health Research Institute, Immunopathology Group, Barakaldo; Ikerbasque, Basque Foundation for Science, Bilbao, Spain

Gerard M.J. Bos
Department of Internal Medicine, Division of Hematology, Maastricht University Medical Center; GROW-School for Oncology and Reproduction, Maastricht University, Maastricht, The Netherlands

Marta Botas
Instituto de Investigaciones Biomédicas Alberto Sols CSIC-UAM, Madrid, Spain

Dirk Brenner
Experimental and Molecular Immunology, Department of Infection and Immunity, Luxembourg Institute of Health; Immunology & Genetics, Luxembourg Centre for Systems Biomedicine, University of Luxembourg, Esch-sur-Alzette, Luxembourg; Odense Research Center for Anaphylaxis (ORCA), Department of Dermatology and Allergy Center, Odense University Hospital, University of Southern Denmark, Odense, Denmark

Magdalena Brkic
TRON - Translational Oncology at the University Medical Center of the Johannes Gutenberg University GmbH, Mainz, Germany

Contributors

Christine E. Brown
T Cell Therapeutics Research Laboratories, Cellular Immunotherapy Center, Department of Hematology and Hematopoietic Cell Transplantation, City of Hope, Duarte, CA, United States

Victor Calvo
Departamento de Bioquímica, Instituto de Investigaciones Biomédicas Alberto Sols CSIC-UAM, Facultad de Medicina, Universidad Autónoma de Madrid, Madrid, Spain

Claudia Cantoni
Department of Experimental Medicine (DIMES), University of Genova; Laboratory of Clinical and Experimental Immunology, IRCCS Istituto Giannina Gaslini, Genoa, Italy

Ondrej Cerny
Institute of Microbiology of the Czech Academy of Sciences, Prague, Czech Republic

Jong Hyun Cho
Department of Pathology, Immunology and Laboratory Medicine, Rutgers University-New Jersey Medical School; Center for Immunity and Inflammation, New Jersey Medical School, Rutgers-The State University of New Jersey, Newark, NJ, United States

Céline Cuche
Institut Pasteur, Université Paris Cité, INSERM-U1224, Unité Biologie Cellulaire des Lymphocytes, Ligue Nationale Contre le Cancer, Équipe Labellisée Ligue-2018, Paris, France

Vincenzo Di Bartolo
Institut Pasteur, Université Paris Cité, INSERM-U1224, Unité Biologie Cellulaire des Lymphocytes, Ligue Nationale Contre le Cancer, Équipe Labellisée Ligue-2018, Paris, France

Catherine Dostert
Experimental and Molecular Immunology, Department of Infection and Immunity, Luxembourg Institute of Health; Immunology & Genetics, Luxembourg Centre for Systems Biomedicine, University of Luxembourg, Esch-sur-Alzette, Luxembourg

Femke A.I. Ehlers
Department of Internal Medicine, Division of Hematology, Maastricht University Medical Center; GROW-School for Oncology and Reproduction, Maastricht University; Department of Transplantation Immunology, Tissue Typing Laboratory, Maastricht University Medical Center, Maastricht, The Netherlands

Sara Fernández-Hermira
Instituto de Investigaciones Biomédicas Alberto Sols CSIC-UAM, Madrid, Spain

Melanie Grusdat
Experimental and Molecular Immunology, Department of Infection and Immunity, Luxembourg Institute of Health; Immunology & Genetics, Luxembourg Centre for Systems Biomedicine, University of Luxembourg, Esch-sur-Alzette, Luxembourg

Manuel Izquierdo
Instituto de Investigaciones Biomédicas Alberto Sols CSIC-UAM, Madrid, Spain

Marie Juzans
Institut Pasteur, Université Paris Cité, INSERM-U1224, Unité Biologie Cellulaire des Lymphocytes, Ligue Nationale Contre le Cancer, Équipe Labellisée Ligue-2018, Paris, France; Department of Pathology and Laboratory Medicine, Children's Hospital of Philadelphia Research Institute, Perelman School of Medicine at the University of Pennsylvania, Philadelphia, PA, United States

Loes F.S. Kooreman
GROW-School for Oncology and Reproduction, Maastricht University; Department of Pathology, Maastricht University Medical Center, Maastricht, The Netherlands

Chiara Lavarello
Core Facilities–Clinical Proteomics and Metabolomics, IRCCS Istituto Giannina Gaslini, Genoa, Italy

Yu Li
Department of Pediatrics, Vagelos College of Physicians and Surgeons, Columbia University Irving Medical Center, New York, NY, United States

Dongfang Liu
Department of Pathology, Immunology and Laboratory Medicine, Rutgers University-New Jersey Medical School; Center for Immunity and Inflammation, New Jersey Medical School, Rutgers-The State University of New Jersey, Newark, NJ, United States

Alireza Naghizadeh
Department of Pathology, Immunology and Laboratory Medicine, Rutgers University-New Jersey Medical School; Center for Immunity and Inflammation, New Jersey Medical School, Rutgers-The State University of New Jersey, Newark, NJ, United States

Jordan S. Orange
Department of Pediatrics, Vagelos College of Physicians and Surgeons, Columbia University Irving Medical Center, New York, NY, United States

Paola Orecchia
IRCCS Ospedale Policlinico San Martino Genova, Genoa, Italy

Monica Parodi
IRCCS Ospedale Policlinico San Martino Genova, Genoa, Italy

Andrea Petretto
Core Facilities–Clinical Proteomics and Metabolomics, IRCCS Istituto Giannina Gaslini, Genoa, Italy

Irene Sanz-Fernández
Instituto de Investigaciones Biomédicas Alberto Sols CSIC-UAM, Madrid, Spain

Ute Schmitt
TRON - Translational Oncology at the University Medical Center of the Johannes Gutenberg University GmbH, Mainz, Germany

Jue Shi
Department of Physics and Department of Biology, Center for Quantitative Systems Biology, Hong Kong Baptist University, Hong Kong, China

Eliana Stanganello
TRON - Translational Oncology at the University Medical Center of the Johannes Gutenberg University GmbH, Mainz, Germany

Lawrence A. Stern
Department of Chemical, Biological and Materials Engineering, University of South Florida, Tampa, FL, United States

Iñigo Terrén
Biocruces Bizkaia Health Research Institute, Immunopathology Group, Barakaldo, Spain

Wei-chung Tsao
Department of Pathology, Immunology and Laboratory Medicine, Rutgers University-New Jersey Medical School; Center for Immunity and Inflammation, New Jersey Medical School, Rutgers-The State University of New Jersey, Newark, NJ, United States

Fulvia Vascotto
TRON - Translational Oncology at the University Medical Center of the Johannes Gutenberg University GmbH, Mainz, Germany

Massimo Vitale
IRCCS Ospedale Policlinico San Martino Genova, Genoa, Italy

Dongrui Wang
T Cell Therapeutics Research Laboratories, Cellular Immunotherapy Center, Department of Hematology and Hematopoietic Cell Transplantation, City of Hope, Duarte, CA, United States

Lotte Wieten
GROW-School for Oncology and Reproduction, Maastricht University; Department of Transplantation Immunology, Tissue Typing Laboratory, Maastricht University Medical Center, Maastricht, The Netherlands

Agata Xella
T Cell Therapeutics Research Laboratories, Cellular Immunotherapy Center, Department of Hematology and Hematopoietic Cell Transplantation, City of Hope, Duarte, CA, United States

Xin Yang
T Cell Therapeutics Research Laboratories, Cellular Immunotherapy Center, Department of Hematology and Hematopoietic Cell Transplantation, City of Hope, Duarte, CA, United States

Steven Zenner
TRON - Translational Oncology at the University Medical Center of the Johannes Gutenberg University GmbH, Mainz, Germany

Yanting Zhu
Department of Physics and Department of Biology, Center for Quantitative Systems Biology, Hong Kong Baptist University, Hong Kong, China

CHAPTER 1

Imaging polarized granule release at the cytotoxic T cell immunological synapse using TIRF microscopy: Control by polarity regulators

Marie Juzans[a,b], Céline Cuche[a], Vincenzo Di Bartolo[a], and Andrés Alcover[a],*

[a]*Institut Pasteur, Université Paris Cité, INSERM-U1224, Unité Biologie Cellulaire des Lymphocytes, Ligue Nationale Contre le Cancer, Équipe Labellisée Ligue-2018, Paris, France*
[b]*Department of Pathology and Laboratory Medicine, Children's Hospital of Philadelphia Research Institute, Perelman School of Medicine at the University of Pennsylvania, Philadelphia, PA, United States*
*Corresponding author: e-mail address: andres.alcover@pasteur.fr

Chapter outline

1	Introduction	2
2	Materials	4
	2.1 Equipment	4
	2.2 Disposable materials	4
	2.3 Chemicals and biological products	4
3	Methods	5
	3.1 Primary C8[+] T cell isolation, differentiation, and lentiviral infection	5
	3.1.1 Peripheral blood mononuclear cell (PBMC) isolation	5
	3.1.2 CD8[+] T cell isolation by magnetic cell sorting	5
	3.1.3 CD8[+] T cell activation and differentiation into CTLs	6
	3.1.4 Assess CTL differentiation by FACS analysis	6
	3.1.5 CD8[+] T cell lentiviral infection to generate silenced CTLs	7
	3.2 Imaging granule secretion by TIRF microscopy	7
	3.2.1 Coat glass-bottom dishes with anti-CD3 antibody	7
	3.2.2 Microscopy	8
	3.3 Quantitative image analysis	9
4	Concluding remarks	11
5	Notes	11

Acknowledgments...12
Declaration of interests..12
References..12

Abstract

Immunological synapse formation results from a profound T cell polarization process that involves the coordinated action of the actin and microtubule cytoskeleton, and the intracellular traffic of several vesicular organelles. T cell polarization is key for both T cell activation leading to T cell proliferation and differentiation, and for T cell effector functions such as polarized secretion of cytokines by helper T cells, or polarized delivery of lytic granules by cytotoxic T cells. Efficient targeting of lytic granules by cytotoxic T cells is a crucial event for the control and elimination of infected or tumor cells. Understanding how lytic granule delivery is regulated and quantifying its efficiency under physiological and pathological conditions may help to improve immune responses against infection and cancer.

Abbreviation

TIRF   total internal reflection fluorescence

1 Introduction

Immunological synapse structure and function depends on a complex process of T cell polarization toward antigen presenting cells or target cells. Polarization is initiated by T cell receptor (TCR) engagement and occurs as the result of the reorganization of both the actin and microtubule cytoskeleton, resulting in the translocation of several organelles to the antigen-presenting cell contact site. These include the Golgi apparatus, and the endosomal and lysosomal compartments that reorient their intracellular vesicle traffic toward the immunological synapse. T cell polarization depends on TCR signal transduction and involves an array of signaling, cytoskeleton and vesicle traffic regulators (Mastrogiovanni, Juzans, Alcover, & Di Bartolo, 2020).

    Cytotoxic T lymphocyte (CTL) effector function depends on the appropriate delivery of lytic granules to the plasma membrane and on their fusion at the zone of contact with target cells. Lytic granules are part of the late endosomal/lysosomal compartment. Molecular motors and intracellular traffic regulators allow lytic granule transport, docking and fusion with the plasma membrane and the delivery of their components, such as perforin and granzymes, to target cells provoking their death (de Saint Basile, Menasche, & Fischer, 2010). Centrosome polarization and docking to the immunological synapse, and actin polymerization and clearance from the

1 Introduction

center of the contact site have been proposed to facilitate the formation of a secretory domain allowing optimal lytic granule delivery and fusion at the CTL-target cell contact (Randzavola et al., 2019; Ritter et al., 2015; Stinchcombe, Bossi, Booth, & Griffiths, 2001; Stinchcombe, Majorovits, Bossi, Fuller, & Griffiths, 2006). However, this seems not to be the sole mechanism (Bertrand et al., 2013; Tamzalit et al., 2020). Conversely, actin recovery at the center of the immunological synapse is associated with termination of lytic granule release (Ritter et al., 2017). Defects in regulators of vesicle traffic or cytoskeleton dynamics are associated with defects in CTL function and immunodeficiency (de Saint Basile et al., 2010).

The role of cell polarity regulators in this process has been recently unveiled (Juzans et al., 2020). Cell polarity regulators are scaffold proteins, endowed with a variety of protein-protein interaction motifs, whose effectors are involved in mechanisms controlling cytoskeleton organization, cell shape and symmetry. Several polarity regulators have been involved in T lymphocyte processes, including cell migration and immunological synapse formation (Mastrogiovanni, Di Bartolo, & Alcover, 2022). Among them, *Discs large homolog 1* (Dlg1) and *Adenomatous polyposis coli* (Apc) play key roles in immunological synapse formation and function in CD4 and CD8 T cells, *via* their control of microtubule network organization at the synapse (Aguera-Gonzalez et al., 2017; Juzans et al., 2020; Lasserre et al., 2010). The protocols we describe here were developed to investigate the involvement of the polarity regulator and tumor suppressor Apc in CTL polarized secretion of lytic granules. In order to have accurate imaging of granule dynamics and fusion, CTLs were induced to polarize on coverslips coated with anti-CD3 antibody. This is a previously described method that generates flat and spread two-dimensional pseudo-immunological synapses on which accurate microscopy imaging may be performed (Bunnell, Barr, Fuller, & Samelson, 2003). We have shown before that these synapse structures clearly display finely organized microtubule networks and centrosome polarization (Aguera-Gonzalez et al., 2017; Juzans et al., 2020). To further ensure accurate visualization of granule docking and fusing at the plasma membrane, we used total internal reflection fluorescence (TIRF) microscopy.

TIRF microscopy uses a specific mode of sample illumination to exclusively excite fluorophores near the adherent cell surface (within ~ 100 nm), corresponding in our case to the plasma membrane at the pseudo synapse. This illumination mode is based on an evanescent wave produced when light rays are totally internally reflected at the interface between the cover glass and the adhered cell plasma membrane (Poulter, Pitkeathly, Smith, & Rappoport, 2015). Since this illumination wave does not propagate deeply into the cell, the region imaged corresponds to the CTL plasma membrane where lytic granules are expected to dock and fuse. It prevents the overlap of images corresponding to granules located deeper inside the cell that would be visualized with other microcopy approaches as scanning or spinning disk confocal microscopy. Therefore, TIRF microscopy allows to differentiate granule polarization defects and docking/fusing defects.

2 Materials
2.1 Equipment
1. Cell culture incubator allowing standard cell culture conditions in a humidified atmosphere (37 °C, 5% CO_2).
2. Standard bench top centrifuge (Eppendorf Centrifuge 5810R or 5415R).
3. Sterile cell culture laminar flow hood safety level II.
4. MACS MultiStand™ (Miltenyi Biotec, No 130-042-303).
5. MidiMACS™ Separator (Miltenyi Biotec, No 130-042-302).
6. Biosafety class 2 laboratory (BSL2 or P2).
7. Microscope: LSM 780 Elyra PS.1 confocal microscope (Zeiss) equipped with a TIRF module and a temperature and CO_2 controlled chamber. Images were acquired with a Plan-Apochromat 100 ×/1.46 numerical aperture oil immersion objective and the ZEN software (Zeiss).
8. ImageJ software with the TrackMate plugin (Tinevez et al., 2017).
9. MACSQuant® Analyzer flow cytometer (Miltenyi Biotec).
10. FlowJo v10 software (FlowJo, LLC) for flow cytometry data analysis.

2.2 Disposable materials
1. 24-well plates for cell culture (Falcon, No 353047).
2. μ-Dish 35 mm, high glass bottom (Ibidi, No 81158 or MatTek No P35G-1.5-10-C) (see **Note 1**).
3. Sterile microcentrifuge tubes (Eppendorf No 3810).
4. 96-well plates for cell culture (TPP, No 92097).

2.3 Chemicals and biological products
1. Lymphocyte Separating Medium Pancoll Human tubes (Pan Biotech, No P04-60125).
2. Magnetic cell sorting CD8+ T cell isolation kit (Miltenyi Biotec, No 130-096-495).
3. LS columns for magnetic cell sorting (Miltenyi Biotec, No 130-042-401).
4. MACS™ MultiStand (Miltenyi Biotec, No 130-042-303).
5. MidiMACS™ Separator (Miltenyi Biotec, No 130-042-302).
6. RPMI 1640 cell culture medium containing GlutaMAX-I and Phenol Red (Gibco, ThermoFisher Scientific, No 61870).
7. Sodium pyruvate (Life Technologies, No 11360).
8. Nonessential amino acids (Life Technologies, No 11140).
9. HEPES (N-(2-Hydroxyethyl)piperazine-N′-(2-ethanesulfonic acid)) (Life Technologies, No 15630-056).
10. Penicillin–streptomycin (Gibco, No 15140-122).
11. Human serum (Dominique Dutscher, No S4190-100).
12. Fetal bovine serum (FBS): HyClone™ SERUM—Research grade fetal bovine serum, origin South America (Dominique Dutscher, No SV30160.03).

13. Purified anti-human CD3ε antibody, clone UCHT1 (BioLegend Inc., No 300402).
14. Anti-human CD28 antibody (Beckman Coulter, No IM1376).
15. Recombinant human IL-2 (PeproTech, No 200-02).
16. Dulbecco's PBS—Modified, w/o $CaCl_2$ and $MgCl_2$ (DPBS), (Gibco, ThermoFisher Scientific, No 14190).
17. Lentiviruses expressing short harpin RNAs (shRNAs), (Sigma-Aldrich) (see **Note 2**).
18. Puromycin (Gibco, No A11138-03).
19. Bovine serum albumin (BSA) (Alpha Diagnostics, No 80400-100).
20. Fixable Viability Stain 450 (BD Biosciences).
21. Antibody cocktail-1 for T cell differentiation: anti-CD3-PE-Cy5 (1/30; clone HIT3a; BD Biosciences), anti-CD8-APC-Cy7 (1/30; clone RPA-T8; Biolegend), anti-CD25-Alexa fluor 488 (1/30; clone M-A251; Biolegend), anti-CD45RA-APC (1/30; cloneHI100; Biolegend), anti-CCR7-PE (1/50; clone G043H7; Biolegend).
22. Saponin (Sigma, No S7900).
23. Antibody cocktail-2 for granzyme B expression: anti-CD3-PE-Cy5 (1/30; clone HIT3a; BD Biosciences), anti-CD8-APC-Cy7 (1/30; clone RPA-T8; Biolegend).
24. Anti-Granzyme-B-PE antibody (1/50; clone GB12; Invitrogen).
25. HCl 1 N (Fluka, No. 84436) solution in 70% ethanol (Fisher Chemical No E/0550DF/21).
26. Poly-L-lysine solution 0.1% (w/v) (Sigma-Aldrich, No P8920).
27. LysoTracker™ Deep Red (Invitrogen, No L12492) to label cytotoxic granules.
28. 5-Carboxyfluorescein diacetate (CFSE) (eBioscience No 65-0850) (see **Note 3**) to label cells.
29. RPMI 1640 medium, no phenol red (Gibco, ThermoFisher Scientific, No 11835).

3 Methods

Fig. 1 shows the graphic summary of the methodology described below.

3.1 Primary C8$^+$ T cell isolation, differentiation, and lentiviral infection

3.1.1 Peripheral blood mononuclear cell (PBMC) isolation

Peripheral blood from healthy donors was obtained from the French National Blood Bank (*Etablissement Français du Sang*) using ethically approved procedures. PBMC were obtained by density gradient centrifugation using Lymphocyte Separating Medium Pancoll Human (see **Note 4**).

3.1.2 CD8$^+$ T cell isolation by magnetic cell sorting

Isolate CD8$^+$ T cells from PBMC by negative selection using magnetic cell sorting using a CD8$^+$ T Cell Isolation Kit and LS columns, following manufacturer's instruction.

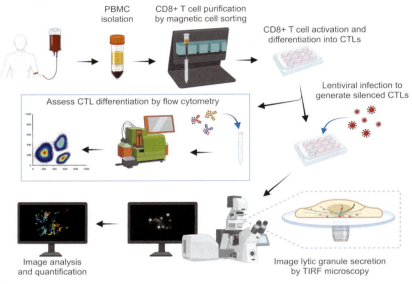

FIG. 1

Graphic summary of the methodology described. Work flow describing the different experimental steps involved in the procedures described in this chapter.

Created with BioRender.com

Culture purified cells in RPMI 1640 cell culture medium supplemented with 10% FBS, 1 mM sodium pyruvate, 1% (v/v) nonessential amino acids, 10 mM HEPES, 0.5% (v/v) penicillin–streptomycin.

3.1.3 CD8$^+$ T cell activation and differentiation into CTLs

1. Coat 24-well plates with 400 µL human anti-CD3 (10 µg/mL) in PBS overnight at 4 °C or 3 h at 37 °C.
2. Wash plates three times with PBS, then incubate 1 h at 37 °C with 500 µL culture medium to prevent non-specific cell binding.
3. Resuspend freshly isolated CD8$^+$ T cells at 2.10^6 cell/mL in culture medium containing 7 µg/mL anti-CD28 (note that anti-CD28 is used soluble, whereas anti-CD3 is coated to the plate) and 100 U/mL recombinant human IL-2, remove medium from plates and distribute 500 µL of cell suspension in each well, incubate plates at 37 °C with 5% CO$_2$ for 2 days (infection) or 6 days (FACS analysis). For FACS analysis, the concentration is adjusted to 2.10^6 cell/mL in culture medium containing 100 U/mL recombinant human IL-2 at day 2 and 4.

3.1.4 Assess CTL differentiation by FACS analysis

1. Take 0.5×10^6 CD8$^+$ T cells at day 0-2-4-6 and place them in a 96 well plate.
2. Wash cells twice with PBS, centrifuge plates for 6 min at $450 \times g$ remove supernatant and add 100 µL of PBS 1× with Fixable Viability Stain 450 (250 ng/mL; BD) diluted 1/1000 (v/v), incubate 10 min at RT.

3. Wash cells twice with PBS, 0.5% (w/v) BSA, spin plates for 6 min at $450 \times g$, remove supernatant and add 75 µL of PBS, 0.5% BSA with antibody cocktail-1 or −2, incubate 30 min at 4 °C.
4. Wash cells twice with PBS, 0.5% BSA, spin plates for 6 min at $450 \times g$, remove supernatant.
5. Fix samples with antibody cocktail-1 by adding 200 µL of PBS, 0.5% BSA, 1% (w/v) paraformaldehyde (PFA), store at 4 °C till flow cytometry analysis (see **point 10**).
6. Fix samples with antibody cocktail-2 by adding 100 µL of 4% PFA, incubate 15 min at RT.
7. Wash cells twice with PBS, 0.5% BSA, spin plates for 6 min at $450 \times g$, remove supernatant and add 60 µL of PBS, 0.5% BSA, 0.05% saponin to permeabilize cell membrane, incubate 10 min at RT.
8. Without wash add 15 µL of anti-Granzyme-B-PE (1/10 v/v; clone GB12; Invitrogen), incubate 30 min at 4 °C.
9. Wash cells twice with PBS, 0.5% BSA, spin plates for 6 min at $450 \times g$, remove supernatant, add 200 µL of PBS, 0.5% BSA, 1% PFA.
10. Analyze samples by flow cytometry. In our case, samples were acquired with a MACSQuant® Analyzer and analyzed using FlowJo v10 software. All samples were gated on forward and side scatter (FSC/SSC), for single cells, and live cells. An example of the FACS gating procedure and typical results are shown in Fig. 2.

3.1.5 CD8⁺ T cell lentiviral infection to generate silenced CTLs

1. After 2-day activation in 24-well plates, infect cells with lentiviruses in a P2 laboratory. Spin plates and remove supernatant, add 900 µL of culture medium in which FBS is replaced by 10% human serum (see **Note 5**), supplemented with 100 U/mL IL-2, add 100 µL of lentivirus suspension (see **Note 6**). In our case, these were coding for control or Apc-specific shRNAs. Incubate at 37 °C with 5% CO_2 for 24 h.
2. Wash cells three times (see **Note 6**) by spinning plates for 7 min at $450 \times g$, remove supernatant, add 1 mL of fresh culture medium. Repeat twice with medium supplemented with 100 U/mL IL-2 and 3.9 µg/mL puromycin. Cells can be taken out of the P2 laboratory. Incubate at 37 °C with 5% CO_2 for 3 days for selection.
3. The night before use, wash cells to remove dead cells. Spin plates for 7 min at $450 \times g$, remove supernatant, add 1 mL of fresh culture medium with 100 U/mL IL-2 without puromycin.

3.2 Imaging granule secretion by TIRF microscopy

3.2.1 Coat glass-bottom dishes with anti-CD3 antibody

1. Wash glass-bottom dishes with HCl-EtOH 70% for 10 min, rinse twice with water and once with EtOH 70% before letting them dry.

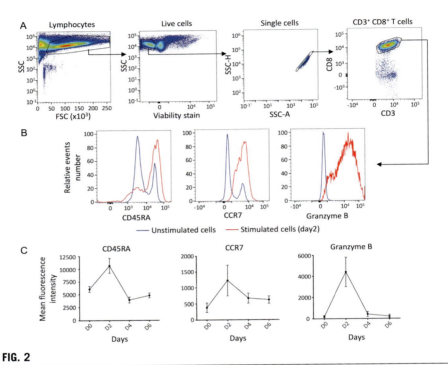

FIG. 2

Expression of T cell molecules characteristic of human CTL differentiation. Expression of CD45RA, CCR7 and granzyme B was analyzed by FACS in CD3$^+$CD8$^+$ T cells stimulated *ex vivo* with CD3+CD28 antibodies for the depicted number of days. At each time point, cells were stained with anti-CD3-PE-Cy5, anti-CD8-APC-Cy7, anti-CD45RA-APC, anti-CCR7-PE and anti-Granzyme-B-PE fluorescent antibodies. (A) FACS gating strategy to measure single, live, CD3$^+$CD8$^+$ T cells. (B) Histogram plots showing fluorescence intensity of CD45RA, CCR7 and granzyme B CTL differentiation markers in CD8$^+$ T cells after 2 days of stimulation. (C) Time course of differentiation marker expression upon CD3+CD28 stimulation. Mean florescence intensity vs time (Mean±SEM, $n=8$ healthy donors).

2. Coat glass-bottom dishes with poly-L-lysine at 0.002% (w/v) in water for 30 min at room temperature, wash once with water, let them dry (see **Note 7**).
3. Coat poly-L-lysine-coated dishes with anti-CD3 antibody at 10 μg/mL in PBS 3 h at 37 °C or overnight at 4 °C, wash three times with PBS, incubate 1 h at 37 °C with culture medium to prevent cell non-specific binding (see **Note 7**).

3.2.2 Microscopy
1. Turn on the TIRF microscope, set up chamber temperature at 37 °C, 5% CO_2 (see **Note 8**). Let the chamber equilibrate for at least 1 h.

2. Collect CTLs by pulling several wells of cells infected with control or silencing lentivirus, wash wells twice with PBS.
3. Wash cells twice with PBS to remove dead cells, spin 7 min at $450 \times g$, resuspend at 1.10^6 cells/mL in PBS with 1 µM CFSE, incubate 5 min at room temperature protected from light.
4. Wash cells twice with PBS, resuspend at 2.10^6 cells/mL in culture medium with 0.1 mM LysoTracker Deep Red, incubate 1 h at 37 °C.
5. Wash cells twice with warm RPMI 1640 without phenol red, resuspend at 1.5×10^6 cells/mL.
6. Equilibrate cells and glass-bottom dishes temperature in the microscope chamber for at least 15 min.
7. Remove medium from one dish, place it on the objective, add 100 µL of cell solution.
8. Follow CTL sedimentation using epifluorescence mode. Once cells reach the dish bottom, switch to TIRF mode to follow their spreading on anti-CD3 antibodies.
9. Start imaging when CTLs are spreading on the surface (pseudo-immunological synapse) (Fig. 2A, middle panel), acquire images in the TIRF plane every 150 ms for 5 min.
10. Repeat steps 7–9 for each image needed. Do not image twice the same dish to prevent acquiring CTLs that have already degranulated.

3.3 Quantitative image analysis

Granule total number and movement are quantified using the TrackMate plugin for ImageJ software (Tinevez et al., 2017), as shown in Fig. 3 and in reference (Juzans et al., 2020). This open-source plugin allows automated particle tracking. Granules are detected in individual cells based on their estimated diameter (0.5 µm approximately). Individual granule position is detected at each time point, allowing the automated linking between positions and missing detection filling. We recommend to set the maximum gap-closing frame on 1 to avoid linking trajectories of two different granules. Generated Excel files provide the number of granules present in the TIRF zone and their LysoTracker mean intensity at each time point. We recommend to consider only strongly fluorescent granules, more likely closer to the plasma membrane. To do so, a threshold of at least $2 \times$ the mean granule fluorescence intensity obtained for an experiment can be set. Granules that are never reaching this threshold are not included in the analysis. Files named Track Statistics also provide tracking duration for each granule, their mean speed during the tracking, and their displacement length (Fig. 3B–D). In addition, files named Spots in Track Statistics provide granule positions (x, y) at each time point allowing to study their directionality. Finally, fusion events are quantified by counting the fluorescence bursts for each cell (Fig. 3E–F).

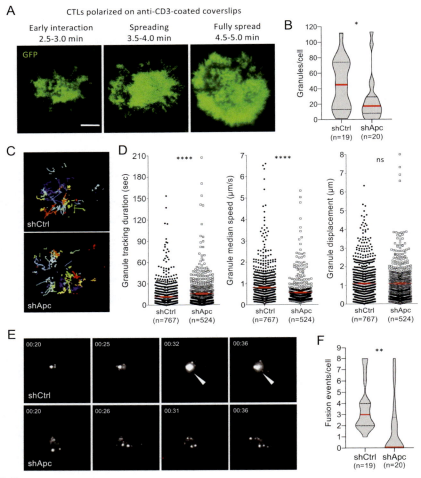

FIG. 3

Quantifying granule secretion by human CTLs using TIRF microscopy. Effect of Apc silencing. *Ex vivo* differentiated primary human CTLs were infected with control or APC shRNA lentiviral vectors expressing GFP and selected with puromycin in culture. Previous to the assay, cells were incubated with LysoTracker to label lytic granules as part of the late endosomal-lysosomal compartment. Cells were set on anti-CD3-coated coverslips to form flat pseudo-immunological synapses and observed by TIRF microscopy. (A) Cell spreading observed by GFP signal in the TIRF zone. (B) Detection of LysoTracker+ granules in the TIRF zone was used as a readout of granule-targeting events per synapse. (C) Individual granule trajectories were obtained by fluorescent tracking of puncta. (D) Tracking duration, average speed, and displacement length were assessed. (E, F) Among the detected lytic granules, some made a fluorescence burst, indicative of fusion with the plasma membrane (arrowhead). The number of bursts was used as a readout of granule fusion events.

Data reported in Juzans, M., Cuche, C., Rose, T., Mastrogiovanni, M., Bochet, P., Di Bartolo, V., et al. (2020). Adenomatous polyposis coli modulates actin and microtubule cytoskeleton at the immunological synapse to tune CTL functions. Immunohorizons, 4, *363–381.*

4 Concluding remarks

The methodology described here allows to measure in a precise and quantitative manner the polarized secretion of granules in human *ex vivo*-differentiated CTLs. It has been applied to compare control and Apc-silenced cells, unveiling a regulatory control of the Apc polarity regulator and tumor suppressor in CTL effector function. It could be therefore applied to the investigation of other genes, or to the comparison of T cells from patients *vs* healthy subjects. Although very precise and quantitative, this methodology only allows the analysis of a restricted number of cells per experiment. This reduces the number of experimental conditions that can be investigated per experiment. Investigation of larger sample panels would need different microscopy set ups.

It is worth noting that we used LysoTracker to monitor lytic granule dynamics. This is a marker of late endosomal and lysosomal compartment, which includes lytic granules, but it is not fully specific for them. Alternative strategies have included the use of fluorescent protein-tagged granzyme B or perforin, both components of lytic granules, see for instance (Pattu et al., 2013; Qu et al., 2011). However, this requires cell infection with lentiviral vectors expressing one of these proteins and would result in their overexpression. Furthermore, full activity or specific localization of these exogenous proteins will not be fully guaranteed.

5 Notes

1. Manipulate glass bottom dishes carefully and avoid putting them directly on the bench. Place them on Kimwipes (Kimtech Science) to avoid scratches on the coverslip.
2. Lentiviruses are produced by transiently transfecting HEK293T cells with the calcium phosphate DNA precipitation technique, as described elsewhere (Kwon & Firestein, 2013). The procedure is based on slow mixing HEPES-buffered saline containing sodium phosphate with a $CaCl_2$ solution containing the DNA. A DNA-calcium phosphate co-precipitate forms, which adheres to the cell surface and is taken up by the cell, presumably by endocytosis. Cells are transfected with pCMV-deltaR8-2, pCMV-env-VSV, and a pLKO.1- puro-CMV-tGFP lentiviral vector expressing or not (as negative control) a shRNA–targeting Apc (5′-GACTGTCCTTTCACCATATT-3′) (Sigma-Aldrich). After 48 h, supernatants are recovered, filtered, and concentrated 40× by ultracentrifugation (26,000 rpm or 11,3000 × g, 1.5 h, 4 °C). Lentivirus stocks are then stored at −80 °C.
3. This reagent is used to visualize cells under the microscope when lentiviral vectors do not contain a fluorescent tag (e.g., GFP). We strongly recommend the use of a fluorescent tag as it allows to identify transduced cells and analyze those with similar fluorescent intensity, reflecting similar lentiviral load.

4. Bring tubes containing the Human Lymphocyte Separating Medium Pancoll to room temperature. Pour carefully 30 mL of blood into each tube. Centrifuge at $800 \times g$ for 30 min in a centrifuge with a swing-out rotor and brake switched off. Harvest the PBMC fraction between the plasma and the Pancoll with a pipette as depicted in the manufacturer's instructions. The membrane present in the tubes prevents contamination with granulocytes and erythrocytes. Wash twice the lymphocytes/PBMCs with DPBS, spin 10 min at $450 \times g$.
5. The use of human serum instead of fetal calf serum strongly increases cell transduction efficiency and cell viability of infected T cells.
6. It is very important not to change plates or even take the cells out during incubation time needed for lentiviral infection, as it will drastically reduce cell transduction efficiency and cell viability. Please note that the biosafety laboratory rules may change with the type of lentivirus used.
7. Glass-bottom dishes can be washed and coated with poly-L-lysine up to a week in advance. However, it is better to coat them with anti-CD3 antibody the day of the experiment.
8. If the microscope is not equipped with a CO_2 chamber, CO_2 independent Leibovitz's L-15 medium, without phenol red (Gibco, ThermoFisher Scientific, No 21083027) supplemented with 2 mg/mL D-glucose may be used.

Acknowledgments

This work was supported by a grant from La Ligue Contre le Cancer Equipe Labellisée Ligue-2018 and Institutional grants from Institut Pasteur and INSERM. MJ has been funded by an Allocation de Recherche Doctorale from La Ligue Nationale Contre le Cancer.

Declaration of interests

The authors have no financial conflict of interest.

References

Aguera-Gonzalez, S., Burton, O. T., Vazquez-Chavez, E., Cuche, C., Herit, F., Bouchet, J., et al. (2017). Adenomatous polyposis coli defines Treg differentiation and anti-inflammatory function through microtubule-mediated NFAT localization. *Cell Reports*, *21*, 181–194.

Bertrand, F., Muller, S., Roh, K. H., Laurent, C., Dupre, L., & Valitutti, S. (2013). An initial and rapid step of lytic granule secretion precedes microtubule organizing center polarization at the cytotoxic T lymphocyte/target cell synapse. *Proceedings of the National Academy of Sciences of the United States of America*, *110*, 6073–6078.

References

Bunnell, S. C., Barr, V. A., Fuller, C. L., & Samelson, L. E. (2003). High-resolution multicolor imaging of dynamic signaling complexes in T cells stimulated by planar substrates. *Science's STKE, 2003*, PL8.

de Saint Basile, G., Menasche, G., & Fischer, A. (2010). Molecular mechanisms of biogenesis and exocytosis of cytotoxic granules. *Nature Reviews. Immunology, 10*, 568–579.

Juzans, M., Cuche, C., Rose, T., Mastrogiovanni, M., Bochet, P., Di Bartolo, V., et al. (2020). Adenomatous polyposis coli modulates actin and microtubule cytoskeleton at the immunological synapse to tune CTL functions. *Immunohorizons, 4*, 363–381.

Kwon, M., & Firestein, B. L. (2013). DNA transfection: Calcium phosphate method. *Methods in Molecular Biology, 1018*, 107–110.

Lasserre, R., Charrin, S., Cuche, C., Danckaert, A., Thoulouze, M. I., de Chaumont, F., et al. (2010). Ezrin tunes T-cell activation by controlling Dlg1 and microtubule positioning at the immunological synapse. *The EMBO Journal, 29*, 2301–2314.

Mastrogiovanni, M., Di Bartolo, V., & Alcover, A. (2022). Cell polarity regulators, multifunctional organizers of lymphocyte activation and function. *Biomedical Journal, 45*, 299–309.

Mastrogiovanni, M., Juzans, M., Alcover, A., & Di Bartolo, V. (2020). Coordinating cytoskeleton and molecular traffic in T cell migration, activation, and effector functions. *Frontiers in Cell and Development Biology, 8*, 591348.

Pattu, V., Halimani, M., Ming, M., Schirra, C., Hahn, U., Bzeih, H., et al. (2013). In the crosshairs: Investigating lytic granules by high-resolution microscopy and electrophysiology. *Frontiers in Immunology, 4*, 411.

Poulter, N. S., Pitkeathly, W. T., Smith, P. J., & Rappoport, J. Z. (2015). The physical basis of total internal reflection fluorescence (TIRF) microscopy and its cellular applications. *Methods in Molecular Biology, 1251*, 1–23.

Qu, B., Pattu, V., Junker, C., Schwarz, E. C., Bhat, S. S., Kummerow, C., et al. (2011). Docking of lytic granules at the immunological synapse in human CTL requires Vti1b-dependent pairing with CD3 endosomes. *Journal of Immunology, 186*, 6894–6904.

Randzavola, L. O., Strege, K., Juzans, M., Asano, Y., Stinchcombe, J. C., Gawden-Bone, C. M., et al. (2019). Loss of ARPC1B impairs cytotoxic T lymphocyte maintenance and cytolytic activity. *The Journal of Clinical Investigation, 129*, 5600–5614.

Ritter, A. T., Asano, Y., Stinchcombe, J. C., Dieckmann, N. M., Chen, B. C., Gawden-Bone, C., et al. (2015). Actin depletion initiates events leading to granule secretion at the immunological synapse. *Immunity, 42*, 864–876.

Ritter, A. T., Kapnick, S. M., Murugesan, S., Schwartzberg, P. L., Griffiths, G. M., & Lippincott-Schwartz, J. (2017). Cortical actin recovery at the immunological synapse leads to termination of lytic granule secretion in cytotoxic T lymphocytes. *Proceedings of the National Academy of Sciences of the United States of America, 114*, E6585–E6594.

Stinchcombe, J. C., Bossi, G., Booth, S., & Griffiths, G. M. (2001). The immunological synapse of CTL contains a secretory domain and membrane bridges. *Immunity, 15*, 751–761.

Stinchcombe, J. C., Majorovits, E., Bossi, G., Fuller, S., & Griffiths, G. M. (2006). Centrosome polarization delivers secretory granules to the immunological synapse. *Nature, 443*, 462–465.

Tamzalit, F., Tran, D., Jin, W., Boyko, V., Bazzi, H., Kepecs, A., et al. (2020). Centrioles control the capacity, but not the specificity, of cytotoxic T cell killing. *Proceedings of the National Academy of Sciences of the United States of America, 117*, 4310–4319.

Tinevez, J. Y., Perry, N., Schindelin, J., Hoopes, G. M., Reynolds, G. D., Laplantine, E., et al. (2017). TrackMate: An open and extensible platform for single-particle tracking. *Methods, 115*, 80–90.

CHAPTER 2

Analysis of centrosomal area actin reorganization and centrosome polarization upon lymphocyte activation at the immunological synapse

Sara Fernández-Hermira[a,†], Irene Sanz-Fernández[a,†], Marta Botas[a,†], Victor Calvo[b], and Manuel Izquierdo[a,*]

[a]Instituto de Investigaciones Biomédicas Alberto Sols CSIC-UAM, Madrid, Spain
[b]Departamento de Bioquímica, Instituto de Investigaciones Biomédicas Alberto Sols CSIC-UAM, Facultad de Medicina, Universidad Autónoma de Madrid, Madrid, Spain
*Corresponding author: e-mail address: mizquierdo@iib.uam.es

Chapter outline

1. Introduction..17
   1.1 The Immunological synapse...17
   1.2 Polarization of centrosome and secretion granules........................18
   1.3 Synaptic actin cytoskeleton regulation of secretory traffic..............18
       1.3.1 Cortical actin cytoskeleton..18
       1.3.2 Centrosomal actin cytoskeleton regulation of secretory traffic.....19
2. Materials, cells, immunological synapse formation and image capture.........20
3. Imaging the immunological synapse...21
   3.1 Measurement of centrosome polarization index (PI).......................21
   3.2 Quantification of F-actin at the centrosomal area.......................23
4. Discussion, future perspectives and concluding remarks......................25
5. Notes...27
Acknowledgments..29

[†]All these authors contributed equally to this manuscipt.

CHAPTER 2 Actin cytoskeleton and polarized traffic

Conflict of interest..29
Author contributions...29
References..29

Abstract

T cell receptor (TCR) and B cell receptor (BCR) stimulation of T and B lymphocytes, by antigen presented on an antigen-presenting cell (APC) induces the formation of the immunological synapse (IS). IS formation is associated with an initial increase in cortical filamentous actin (F-actin) at the IS, followed by a decrease in F-actin density at the central region of the IS, which contains the secretory domain. This is followed by the convergence of secretion vesicles towards the centrosome, and the polarization of the centrosome to the IS. These reversible, cortical actin cytoskeleton reorganization processes occur during lytic granule secretion in cytotoxic T lymphocytes (CTL) and natural killer (NK) cells, proteolytic granules secretion in B lymphocytes and during cytokine-containing vesicle secretion in T-helper (Th) lymphocytes. In addition, several findings obtained in T and B lymphocytes forming IS show that actin cytoskeleton reorganization also occurs at the centrosomal area. F-actin reduction at the centrosomal area appears to be associated with centrosome polarization. In this chapter we deal with the analysis of centrosomal area F-actin reorganization, as well as the centrosome polarization analysis toward the IS.

Abbreviations

AIP	average intensity projection
APC	antigen-presenting cell
BCR	B-cell receptor for antigen
c	center of mass
CMAC	CellTracker™ Blue (7-amino-4-chloromethylcoumarin)
cSMAC	central supramolecular activation cluster
CTL	cytotoxic T lymphocytes
Dia1	diaphanous-1
dSMAC	distal supramolecular activation cluster
F-actin	filamentous actin
FCS	fetal calf serum
FMNL1	formin-like 1
GFP	green fluorescent protein
IS	immunological synapse
LLSM	lattice light-sheet microscopy
MFI	mean fluorescence intensity
MHC	major histocompatibility complex
MIP	maximal intensity projection

MTOC	microtubule-organizing center
MVB	multivesicular bodies
NK	natural killer
PI	polarization index
PKCδ	protein kinase C δ isoform
pSMAC	peripheral supramolecular activation cluster
ROI	region of interest
RT	room temperature
SEE	Staphylococcal enterotoxin E
SMAC	supramolecular activation cluster
TCR	T-cell receptor for antigen
Th	T-helper
TIRFM	total internal reflection microscopy
TRANS	transmittance

1 Introduction
1.1 The Immunological synapse

T and B lymphocyte activation by APC takes place at a specialized cell to cell interface called the IS. APC have the ability to present antigens to T lymphocytes bound to major histocompatibility complex (MHC) molecules (Fooksman et al., 2010; Huppa & Davis, 2003). In contrast, B lymphocytes can directly recognize antigen tethered to the cell surface of specialized APC (Yuseff, Pierobon, Reversat, & Lennon-Dumenil, 2013). In addition, NK cells were first noticed for their ability to kill tumor cells after IS formation, but without any priming or prior activation, in contrast to CTL, which need priming and activation by APC (Lagrue et al., 2013). IS establishment by T and B lymphocytes and NK cells is a very dynamic, plastic and critical event, acting as a tunable signaling platform that integrates spatial, mechanical and biochemical signals, involved in specific, cellular and humoral immune responses (De La Roche, Asano, & Griffiths, 2016; Fooksman et al., 2010). The general architecture of the IS is described by the formation of a concentric, bullseye spatial pattern, named the supramolecular activation complex (SMAC), upon cortical actin reorganization (Billadeau, Nolz, & Gomez, 2007; Carisey, Mace, Saeed, Davis, & Orange, 2018; Griffiths, Tsun, & Stinchcombe, 2010; Kuokkanen, Sustar, & Mattila, 2015; Yuseff et al., 2013). In the IS made by T and B lymphocytes, this reorganization yields a central cluster of antigen receptors bound to antigen called central SMAC (cSMAC) produced by centripetal traffic, and a surrounding adhesion molecule-rich ring, called peripheral SMAC (pSMAC), which appears to be crucial for adhesion with the APC (Fooksman et al., 2010; Monks, Freiberg, Kupfer, Sciaky, & Kupfer, 1998). The distal SMAC (dSMAC) is located surrounding the T and B lymphocytes pSMAC, at the edge of the contact area with the APC, and comprises a circular array of dense F-actin (Griffiths et al., 2010; Le Floc'h & Huse, 2015; Rak, Mace, Banerjee, Svitkina, & Orange, 2011; Ritter, Angus, & Griffiths, 2013). More recently, several super-resolution imaging techniques have revealed that, upon TCR–antigen interaction, at least four discrete F-actin networks form and maintain the shape and function of this canonical IS (Blumenthal & Burkhardt, 2020; Hammer, Wang, Saeed, & Pedrosa, 2018).

1.2 Polarization of centrosome and secretion granules

IS formation induces the convergence of T and B lymphocytes and NK cells secretion vesicles toward the centrosome and, simultaneously, the polarization of the centrosome (the major microtubule-organizing center—MTOC—in lymphocytes) toward the IS (De La Roche et al., 2016; Huse, 2012). These traffic events, acting together, lead to polarized secretion of extracellular vesicles and exosomes coming from multivesicular bodies (MVB) in T and B lymphocytes (Alonso et al., 2011; Calvo & Izquierdo, 2020; Herranz et al., 2019; Mazzeo, Calvo, Alonso, Mérida, & Izquierdo, 2016; Peters et al., 1991) lytic granules in CTL and NKs (De La Roche et al., 2016; Lagrue et al., 2013; Stinchcombe, Majorovits, Bossi, Fuller, & Griffiths, 2006) stimulatory cytokines in Th cells (Huse, Quann, & Davis, 2008) and lytic proteases in B lymphocytes (Yuseff et al., 2013).

1.3 Synaptic actin cytoskeleton regulation of secretory traffic
1.3.1 Cortical actin cytoskeleton

Actin cytoskeleton reorganization plays a central role in IS maintenance, but also in antigen receptor-derived signaling in T and B lymphocytes (Billadeau et al., 2007) and activating receptor signaling in NK cells (Ben-Shmuel, Sabag, Biber, & Barda-Saad, 2021). Please refer to the excellent reviews on this subject in the IS made by B lymphocytes (Yuseff et al., 2013; Yuseff, Lankar, & Lennon-Dumenil, 2009), T lymphocytes (Billadeau et al., 2007; Ritter et al., 2013), and NK cells (Ben-Shmuel et al., 2021; Lagrue et al., 2013). The concentric F-actin architecture of the IS and the actin cortical cytoskeleton reorganization are shared by B lymphocytes, CD4[+] Th lymphocytes, CD8[+] CTL, and NK cells (Brown et al., 2011; Le Floc'h & Huse, 2015). Remarkably, all these immune cells exhibit the ability to form synapses and to directionally secrete proteases, cytokines or cytotoxic factors at the IS. Thus, this polarized secretion in the context of the F-actin synaptic architecture, most probably, enhances the specificity and the efficacy of the subsequent responses to these factors (Le Floc'h & Huse, 2015) by spatially and temporally focusing the secretion at the synaptic cleft (Billadeau et al., 2007), which avoids the stimulation or death of bystander cells.

At the early phases of IS formation, F-actin accumulates at the contact area (called IS interface) of the immune cell with the APC, generating filopodia and lamellipodia, that produce dynamic changes between extension and contraction in the lymphocyte over the surface of the APC (Le Floc'h & Huse, 2015). Subsequently, once IS evolution has stabilized, F-actin reduction from the cSMAC appears to facilitate secretion toward the APC by focusing secretion vesicles on the IS (Stinchcombe et al., 2006) and, almost simultaneously, cortical F-actin accumulates into the dSMAC. Thus, F-actin forms a permissive network at the IS of CTL and NK cells (Carisey et al., 2018; Ritter et al., 2015). These reversible, cortical actin cytoskeleton reorganization processes occur during lytic granules secretion by both CTL and NK cells, but also during the polarization of some cytokine-containing secretion vesicles in Th lymphocytes (Chemin et al., 2012; Griffiths et al., 2010; Ritter et al., 2015), despite both the nature and cargo of the secretion vesicles in these cell types being quite different. Moreover, it has been shown that cortical F-actin density recovery at the IS leads to termination of lytic granule secretion in CTL (Ritter et al., 2017), supporting the role of actin cytoskeleton in initiation, but also in termination of granule secretion.

1.3.2 Centrosomal actin cytoskeleton regulation of secretory traffic

F-actin reduction at the cSMAC does not simply allow secretion, as it apparently plays an active role in the initiation of centrosome and secretory granules movement toward the IS (Ritter et al., 2015; Stinchcombe et al., 2006). In this context, some results suggest that cortical actin reorganization at the IS is necessary and sufficient for centrosome and lytic or cytokine-containing granules polarization (Chemin et al., 2012; Ritter et al., 2015; Sanchez, Liu, & Huse, 2019).

However, other results show that the elimination of certain F-actin regulators such as formins FMNL1 or Dia1 inhibits centrosome polarization without affecting Arp2/3-dependent cortical actin reorganization (Gomez et al., 2007) supporting that, at least in the absence of FMNL1 or Dia1, cortical actin reorganization is not sufficient for centrosome polarization. Conversely, in the absence of cortical actin reorganization at the IS occurring in Jurkat T lymphocytes lacking Arp2/3, the centrosome can polarize normally to the IS (Gomez et al., 2007; Kumari, Curado, Mayya, & Dustin, 2014). All these results support that centrosome and secretory granules polarization induced by IS formation are regulated by HS1/WASp/Arp2/3-dependent cortical and formin-dependent non-cortical actin networks (Gomez et al., 2007; Kumari et al., 2014). Thus, analysis of all these F-actin networks at different subcellular locations is necessary to achieve the full picture of the cellular actin reorganization processes leading to polarized secretion. Included among non-cortical actin networks, centrosomal area F-actin depletion appears to be crucial to allow centrosome polarization towards the IS in B lymphocytes stimulated with BCR-ligand-coated beads, used as a B lymphocyte synapse model (Obino et al., 2016). These data do not directly allow inferring either the sufficiency or the relative contribution of each F-actin network (cortical and centrosomal area) to centrosome polarization, unless specific experimental approaches are designed to address this point (Bello-Gamboa et al., 2020). The above-mentioned results on B lymphocytes, together with our results showing that PKCδ interference affects cortical F-actin at the IS (Herranz et al., 2019), prompted us to study the involvement of centrosomal area F-actin in centrosome polarization in cell-to-cell synapses made by T lymphocytes and its potential regulation by PKCδ (Bello-Gamboa et al., 2020). The experimental approach described here allows the simultaneous evaluation of centrosome polarization and centrosomal area F-actin, at the single cell level. To develop these analyses, modifications of the methods previously described in

B lymphocytes to measure normalized centrosome polarization (Obino et al., 2017; Saez et al., 2019) and centrosomal area F-actin (Ibanez-Vega, Del Valle Batalla, Saez, Soza, & Yuseff, 2019; Obino et al., 2016) were used. Following this approach, we have shown that, upon Th lymphocyte IS formation, centrosomal area F-actin decreased concomitantly with centrosome polarization to the IS, and a linear correlation between these two parameters exists (Bello-Gamboa et al., 2020).

2 Materials, cells, immunological synapse formation and image capture

1. Raji B and Jurkat T (clone JE6.1) cell lines are obtained from the ATCC.
2. Cell lines are cultured in RPMI 1640 medium containing L-glutamine (Invitrogen) with 10% heat-inactivated fetal calf serum (FCS) (Gibco) and penicillin/streptomycin (Gibco) and 10mM HEPES (Lonza).
3. For IS formation, Raji cells are attached to Ibidi 8 microwell culture dishes (IBIDI) (glass bottom, for acetone-fixation) using poly-L-lysine (20μg/mL, 1h incubation at 37°C, SIGMA), labeled with CellTracker™ Blue (7-amino-4-chloromethylcoumarin) (CMAC, 10μM, 45min incubation at 37°C, ThermoFisher) and, after CMAC washing, are pulsed with 1μg/mL Staphylococcal enterotoxin E (SEE, 45min incubation at 37°C, Toxin Technology, Inc). After careful aspiration of the culture medium containing SEE, Jurkat cells are directly added to the microwells, so that IS are formed. CMAC labeling allows to discriminate Jurkat/Raji conjugates. Please refer to the following references for further details, since this IS model has been exhaustively described (Alonso et al., 2011; Bello-Gamboa et al., 2019; Montoya et al., 2002). After 30min to 1h of cell conjugate formation, end-point fixation is performed with chilled acetone for 5min (see *Note 1*).
4. Perform immunofluorescence following standard protocols. Blocking and permeabilization solutions for sequential antibody/phalloidin incubations and washing steps (3 ×) after each sequential incubation contain saponin (0.1%). The centrosome is labeled with mouse monoclonal anti-γ-tubulin (1/2000 dilution in permeabilization solution, 45min at room temperature-RT-, Clone GTU-88, SIGMA) and an appropriate secondary antibody coupled to AF546 (1/200 dilution in permeabilization solution, 30min at RT, Thermofisher). F-actin is labeled with phalloidin AF647 (1/100 dilution in permeabilization solution, 30min at RT, Thermofisher). CMAC fluorescence does not overlap with these fluorochromes.
5. Perform image capture with confocal microscope. Around 50 optical sections are acquired using a 0.2–0.3μm Z-step size to have an appropriate axial resolution. Confocal settings we have commonly used are (Leica SP8 confocal microscope): Objective 100x_oil 1.4 NA

Zoom_4
Scan Velocity_700Hz_Bidirectional
1024 × 1024
Pixel Size_0.028 μm
Image Size_29.06 μm × 29.06 μm
Pinhole_0.7
Optical section_0.704 μm
ZStepSize_0.3 μm
Z-stack_variable (*top* and *bottom*)
8 bit
Frame Average-2
Line Average-1

3 Imaging the immunological synapse
3.1 Measurement of centrosome polarization index (PI)

1. Open the confocal file in the public, multiplatform software ImageJ or FIJI (https://imagej.nih.gov/ij/) using the "Bio-Formats" import plugin (see *Notes 2 and 3*). Save separately the stacks of the different channels in TIF format.
2. Draw the regions of interest (ROI) as seen in Fig. 1 as previously described (Bello-Gamboa et al., 2020; Obino et al., 2017). Saving all the ROIs by using the "ROI Manager" Image J submenu will allow to recover the ROIs when required (see *Note 4*).
    2.1. Cell shape and cell center of mass.
    - Make the maximal intensity projection (MIP) with the "Z" stacks of the appropriate channel (i.e., phalloidin channel) where cell shape is properly visualized thanks to the cortical F-actin labeling, by selecting "Image" > "Stacks" > "Z Project" > "Max Intensity." Draw the cell outline ("freehand" selection in ImageJ toolbar) in the MIP (see *Note 5*). Save the cell outline as a ROI (cell ROI).
    - Press "Measure" in order to obtain the value of the cell center of mass (cell^C) (XM and YM coordinates) (see *Note 6*).
    - Draw a small rectangle ("rectangular" selection in ImageJ toolbar) inside the cell. Press "Edit" > "Selection" > "Specify" and introduce the values obtained from the cell^C in the X and Y coordinates. Select "Scaled units (microns)." Save it as a ROI (cell^C ROI).
    2.2. Analysis of "B" distance.
    - Make a MIP by selecting "Image" > "Stacks" > "Z Project" > "Max Intensity" with the "Z" stacks of the γ-tubulin channel where the centrosome is visualized and establish the centrosome center of mass (centrosome^C).

CHAPTER 2 Actin cytoskeleton and polarized traffic

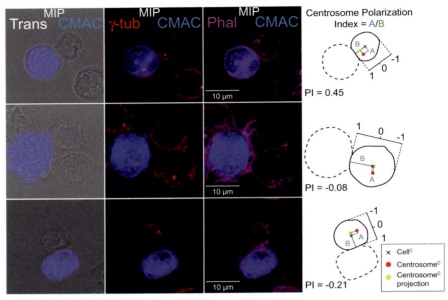

FIG. 1

Measurement of centrosome polarization index. Jurkat cells were challenged with CMAC-labeled (blue), SEE-pulsed Raji cells for 1 h, fixed, stained with anti-γ-tubulin AF546 to label the centrosome (red) and phalloidin AF647 to label F-actin (magenta) and imaged by confocal fluorescence microscopy. In the left panels, Maximal Intensity Projection (MIP) of the indicated merged channels (transmittance-Trans-+CMAC, γ-tubulin+CMAC and phalloidin+CMAC) are shown. First row shows an example of a centrosome polarized towards the IS, and second row displays a cell exhibiting a non-polarized centrosome while the lower images show a centrosome in the opposite direction of the IS. The diagrams to the right show the ROIs and the measurements needed for centrosome PI determination: black crosses label cell[C]; red dots indicate centrosome[C]; yellow dots represent intersections between the B and the projection segments and the segments represent the distances ("A", blue; "B", green) used for centrosome PI calculation (A/B). Note that if the intersection between the B and the projection segments is between the cell[C] and the IS, the value of distance "A" will be positive; otherwise, it must be considered negative. Thus, centrosome PI will vary between +1 (completely polarized) and −1 (completely anti-polarized). Raji and Jurkat cells are labeled with discontinuous and continuous white lines, respectively. (For interpretation of the references to color in this figure legend, the reader is referred to the web version of this article.)

- On the resulting MIP, draw the B segment as the shortest connection between cell[C] and the IS ("straight line" selection in ImageJ toolbar). Save it as a ROI (B ROI).
- Press "Measure" to determine "B" segment length ("B distance").

2.3. Analysis of "A" value.
- Draw the projection segment ("straight line" selection in ImageJ toolbar), from the centrosome^C perpendicular to the axis defined by the B segment. Save it as a ROI (projection segment ROI).
- Draw the A segment ("straight line" selection in ImageJ toolbar) from the cell^C to the intersection between the B and the projection segments. Save it as a ROI (A ROI).
- Press "Measure" to determine "A" length. If the centrosome^C projection is located between the cell^C and the IS, this value will stay positive. However, it must be considered negative if the centrosome^C projection is on the opposite side of the cell.

3. Calculate the centrosome PI by dividing "A" distance by "B" distance, thus it will be normalized by the distance between the cell^C and the IS (Fig. 1). The result will vary between +1 (completely polarized) and −1 (completely anti-polarized) (Fig. 1).

3.2 Quantification of F-actin at the centrosomal area

1. Define the position of the centrosome and the cellular optical sections to be studied.

   1.1. Open the confocal image capture file in the confocal visualization program (see *Note 7*).

   1.2. Determine the "Z" substack in which the centrosome is visually displayed. The appropriate volume to quantify the amount of F-actin around the centrosome is a 2 μm height cylinder with a 2 μm diameter circular base, as previously described (Bello-Gamboa et al., 2020; Ibanez-Vega et al., 2019). Therefore, adjust the "Z" substack selection to this 2 μm height. After manual selection of the Z optical section containing the maximal signal corresponding to the centrosome (γ-tubulin signal), a 2 μm-high substack of the F-actin channel centered at the centrosome^C is defined. Write down the initial and final optical slice. This "centrosomal substack" will comprise 10 optical sections (Z step size = 0.2 μm).

   1.3. Determine the "Z" substack in which all the cell is visually displayed, using phalloidin staining as a reference (see *Notes 8 and 9*). Annotate the initial and final optical slice. This will be considered as the "cell substack."

2. Quantification of centrosomal area F-actin Mean Fluorescence Intensity (MFI) ("C").

   2.1. Import the confocal image capture file to ImageJ by using the "Bio-Formats" plugin (see *Notes 2 and 3*) and visualize the 2 μm-high centrosomal substack of the F-actin channel, as defined in step 1.2.

   2.2. Draw on the file a 2 μm-diameter oval ROI (using "oval" selection in ImageJ toolbar). Then, select "Edit" > "Selection" > "Specify" and choose "Scaled units (microns)." Adjust ROI length and width to 2 μm and place the circular ROI so its center overlaps the centrosome^C (see *Note 10*) (Fig. 2). Save it as the centrosomal area ROI (see *Note 4*).

CHAPTER 2 Actin cytoskeleton and polarized traffic

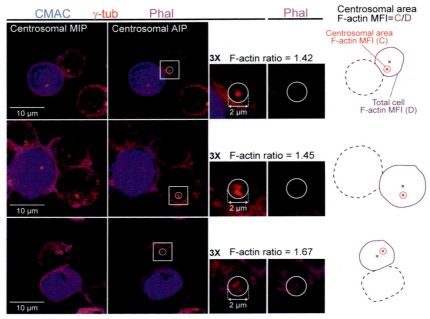

FIG. 2

Quantification of centrosomal area F-actin. Jurkat cells were challenged with CMAC-labeled (blue), SEE-pulsed Raji cells for 1 h, then they were fixed, stained with anti-γ-tubulin AF546 to label the centrosome (red) and phalloidin AF647 to label F-actin (magenta) and imaged by confocal fluorescence microscopy. In the left panels, Maximal and Average Intensity Projections (MIP and AIP, respectively) corresponding to the 2 μm-high, centrosomal substack are shown of the indicated merged channels (CMAC + γ-tubulin + phalloidin) of three representative cells: upper images show a centrosome polarized towards the IS, middle and lower images show centrosomes in the opposite direction of the IS. In the middle panels, 3× enlargements of the centrosomal areas for the three merged channels (left) and for the phalloidin channel (right), defined by white squares in the AIPs, are shown. The 2 μm-diameter ROIs used to calculate centrosomal area F-actin MFI are indicated as white circles. The diagrams to the right indicate the ROIs used in the quantification of the MFI ratio: black crosses, cell[C]; red dots, centrosome[C]; 2 μm-diameter red circles where centrosomal area F-actin MFI ("C") is measured; purple cell outlines where total cell F-actin MFI ("D") is calculated. Centrosomal area F-actin MFI ratio is calculated as C/D, and the corresponding centrosome PI values are shown. Raji and Jurkat cells are drawn using discontinuous and continuous lines, respectively. (For interpretation of the references to color in this figure legend, the reader is referred to the web version of this article.)

2.3. Make an Average Intensity Projection (AIP) of the centrosomal substack of the F-actin channel: "Image" > "Stacks" > "Z Project" > "Average Intensity", by selecting the optical slices defined in step 1.2. Thus, the 2 μm-high "Z" centrosomal substack established in step 1.2 will be delimited. The F-actin AIP of the centrosomal substack is generated (centrosomal F-actin AIP).

2.4. Select the centrosomal area ROI centered at the centrosome[C] from step 2.2 in the centrosomal F-actin AIP from step 2.3, and delete the outside of the ROI by selecting: "Edit" > "Clear Outside."

2.5. Perform an automatic threshold: "Image" > "Adjust" > "Threshold" (see *Note 11*).

2.6. In order to measure the MFI, select "Analyze" > "Measure" (see *Note 12*). The thresholded, centrosomal area F-actin MFI in the centrosomal area ROI is calculated (F-actin MFI in ROI "C", Fig. 2).

3. Quantification of total cell F-actin MFI ("D").

   3.1. Make a MIP of the F-actin channel: "Image" > "Stacks" > "Z Project" > "Max Intensity", by selecting the optical sections defined in step 1.3 (see *Note 13*).

   3.2. Draw the outline of the cell on this MIP ("freehand" selection in ImageJ toolbar) and save it as the cell ROI (see *Notes 4, 5, 14 and 15*) (Fig. 2).

   3.3. Make an AIP of the cell substack of the F-actin channel: "Image" > "Stacks" > "Z Project" > "Average Intensity" by selecting the optical sections defined in step 1.3 (see *Note 13*).

   3.4. Select the cell ROI from step 3.2 on this AIP and delete the outside of the ROI by selecting "Edit" > "Clear Outside."

   3.5. Perform an automatic threshold: "Image" > "Adjust" > "Threshold" (see *Note 11*).

   3.6. In order to measure the MFI, select "Analyze" > "Measure" (see *Note 16*). The thresholded, total cell F-actin MFI in the cell ROI is calculated (F-actin MFI in ROI "D", Fig. 2).

4. Determine the centrosomal area F-actin MFI ratio by dividing "C" by "D" (Fig. 2). Thus, centrosomal F-actin measurements will be normalized among samples from different days and different F-actin intensities.

4 Discussion, future perspectives and concluding remarks

The method described above is not automated and, therefore, to perform the quantification the user must select single cells forming IS one by one, which may make it difficult to evaluate results in a fully unbiased form and also may be time-consuming to perform for an untrained user. However, a trained user may analyze up to 20–30 synapses per hour. To prevent biases, please try to randomly select as many synaptic conjugates from different fields as possible. In addition, we have tried artificial intelligence programs such as the NisAR supplement Ai (NIKON) to automatically identify and select synaptic conjugates for subsequent analysis. Unfortunately, in our experience this approach cannot distinguish canonic IS from irrelevant cell contacts or cell aggregates randomly produced by the high cell concentrations used to favor IS formation (Bello-Gamboa et al., 2019). Improvement of the artificial intelligence algorithms will contribute to solve this problem. Evaluation of the cup-shape profile of the effector T lymphocyte and/or lamellipodium formation observed in the transmittance channel may help to identify IS. A criterion we recommend to unambiguously select canonic IS and distinguish them from irrelevant cell contacts is to use the F-actin probe phalloidin (Figs. 1 and 2), since it is known that F-actin accumulates at the lamellipodium in the synaptic contact area (please refer to Section 1). CMAC labeling facilitates identification of Jurkat/Raji conjugates.

In order to somewhat automate image analyses and save time we have used the ImageJ "Macro" language (IJM), that is a scripting language built into ImageJ that allows controlling many aspects of this software. A macro is a simple program that automates a series of ImageJ commands and may facilitate dealing with image stacks. Programs written in the IJM, or macros, can be used to perform sequences of actions in a fashion expressed by the programs design. The easiest way to create a macro in ImageJ is to record a series of commands using the command recorder ("Plugins" > "Macros" > "Record"). A macro is saved as a text file and executed by selecting a menu command, by pressing a key or by clicking on an icon in the ImageJ toolbar. Please refer to the excellent ImageJ tutorials https://imagej.nih.gov/ij/developer/macro/macros.html. We have found the following macros in txt format very useful:

run("Make Substack…","slices=X-Y);
run("Z Project…", "projection=[Average Intensity]");

The first macro simplifies substack generation (changing X and Y characters by the optical section number) and the second macro directly generates the AIP from the former substack or a different substack in a single operation. Thus, these macros could be used in steps 2.3 and 3.3, and can be used separately or together, as required.

Several authors, including ourselves, operationally used the AIP of a 2 μm-high, F-actin confocal substack centered at the centrosome on an arbitrarily-defined, 2 μm-diameter circular region of interest (ROI) centered at the centrosome, to measure centrosomal area F-actin reorganization triggered by Th (Bello-Gamboa et al., 2019) and B lymphocytes (Ibanez-Vega et al., 2019; Obino et al., 2016) IS formation. However, it cannot be excluded that other organelles, such as the Golgi (Colon-Franco, Gomez, & Billadeau, 2011) or endosomes/MVB (Calabia-Linares et al., 2011; Ueda, Morphew, Mcintosh, & Davis, 2011), that are competent in reorganizing F-actin or tubulin cytoskeleton, may be included in this area, and thus we cannot exclude the contribution of Golgi/MVB to the F-actin reorganization at the centrosomal area during centrosome polarization. The imaging techniques used in these reports (confocal microscopy, total internal reflection microscopy—TIRFM-, epifluorescence microscopy plus deconvolution) do not provide enough resolution to discriminate between F-actin assembly at the pericentrosomal matrix or other membrane-bound organelles, that may be included in the 2 μm-diameter centrosomal area ROI (Bello-Gamboa et al., 2019; Ibanez-Vega et al., 2019; Obino et al., 2016). Indeed, in the future, emerging and promising imaging techniques applied to living cells, harboring high spatio-temporal resolution such as lattice light-sheet microscopy (LLSM) (Fritzsche et al., 2017; Ritter et al., 2015), combined with non-diffraction limited, super-resolution microscopy (Calvo & Izquierdo, 2018; Fritzsche et al., 2017), may contribute to a better definition of centrosomal area F-actin structure and function, as it occurred for the four distinct synaptic F-actin networks in the canonical IS (Blumenthal & Burkhardt, 2020; Hammer et al., 2018). By using these approaches, it will be interesting to analyze, for instance, whether centrosomal F-actin clearing that occurs during centrosome polarization to the IS is a reversible event, as it occurs with cortical F-actin in IS made by CTL (Ritter et al., 2015, 2017). In the future, this analysis will allow extending the knowledge regarding the contribution of centrosomal area F-actin to polarized secretion obtained in B and Th lymphocytes IS (Obino et al., 2016; Bello-Gamboa et al., 2019, 2020) to CTL and NK cells IS, due to the similarities existing among all these IS. In addition, all these approaches together with our analysis will enable the study of the contribution of centrosomal area F-actin to polarized secretion during directional and invasive cell migration, both of lymphoid and non-lymphoid cells (Calvo & Izquierdo, 2021). This would clarify whether centrosomal area F-actin reorganization occurs only during IS formation or contributes also to other biologically relevant cellular polarization processes. This analysis will enable to address how the distinct cortical and centrosomal F-actin networks are regulated, how these networks integrate into cell surface receptor-evoked signaling networks, as well as their interconnections with tubulin cytoskeleton, that constitute an intriguing and challenging biological issue (Calvo & Izquierdo, 2021; Dogterom & Koenderink, 2019).

5 Notes

1. Acetone precipitates FCS protein, thus a previous wash in warm culture medium without FCS prior to fixation is recommended. Acetone (and acetone vapor) dissolves plastic, thus keep the microwell with fixative on ice at 4 °C, and the microwell plastic lid should be removed. Recommended maximal fixation time is 5 min, since longer times dehydrate samples in excess (producing cell shrinking and cell shape changes). Acetone fixation does not allow so clean staining of centrosome with anti-γ-tubulin as methanol fixation does, although centrosome staining is still evident (Figs. 1 and 2). We noticed this occurs regardless of blocking and permabilization conditions and the number of washings steps during immunofluorescence. Paraformaldehyde or glutaraldehyde fixation does not allow staining of centrosome with anti-γ-tubulin. An important issue is that F-actin staining with phalloidin is not compatible with

methanol fixation, whereas it is compatible with acetone fixation. Thus, we have chosen acetone fixation since it constitutes a good compromise to circumvent these caveats (Abrahamsen, Sundvold-Gjerstad, Habtamu, Bogen, & Spurkland, 2018). As an alternative, some authors have used glutaraldehyde fixation instead, which enables proper F-actin labeling, but not anti-γ-tubulin staining (Ibanez-Vega et al., 2019; Obino et al., 2016). In this condition, centrosome position was indirectly inferred with anti-α-tubulin staining, as the brightest point where the microtubules converge by using an appropriate threshold. We believe our approach is more direct indeed.

2. If a channel is not properly visualized, you can modify its optical parameters in "Image" > "Adjust" > "Brightness/Contrast." Please do not apply these changes especially during F-actin measurement; otherwise, intensity values will be modified.
3. In order to improve and synchronize the visualization of the different channels you can select "Window" > "Tile" and "Analyze" > "Tools" > "Synchronize windows." Also, you can visualize and merge the channels you are interested in by selecting "Image" > "Color" > "Merge Channels."
4. To open the ROI manager press "Analyze" > "Tools" > "ROI manager." Select "Show All" so afterwards you can always recover and visualize the previously saved ROIs. Leave this window open for next steps.
5. In ImageJ the drawing is performed with the "freehand" tool using a proper channel to outline the cell (i.e., phalloidin). Other analysis software includes an "autodetect" option. For the study of the IS this option is useful only if the effector cell being studied has a clear differential staining (e.g., expresses a reporter gene such as GFP).
6. The first time the analysis is performed, it is necessary to establish the parameters that are going to be measured: "Analyze" > "Set Measurements" and select "Center of mass."
7. Always use the same computer screen in order to avoid discrepancies in ROI drawing and Z substack selection and minimize the error between measurements.
8. If cell shape differs considerably from one "Z" slice to another, divide the "Z" slices into different substack groups.
9. In order to avoid discrepancies and minimize the error between measurements, do not select "Z" stacks where the F-actin is barely seen.
10. The circular ROI must not include the IS F-actin, which may happen in cells with a highly polarized centrosome towards the IS. Then the centrosomal area may overlie the synaptic membrane rich in cortical F-actin and bias the measurements.
11. Use "Default" set-up and do not select "Apply," since it would affect the MFI measurements.
12. The first time the analysis is performed, it is necessary to establish the parameters that are going to be measured: "Analyze" > "Set Measurements" and select "Limit to threshold," "Area" and "Mean gray value."

13. If different substack groups were selected in step 1.3, make a MIP for each substack.
14. Be careful not to include F-actin from nearby cells.
15. Draw a ROI for each MIP of step 3.1.
16. If different substacks were selected in 1.3, MFI value must be computed by calculating a weighted average of the values obtained by following steps 3.2–3.6 in the different substack groups and considering the number of "Z" slices in each substack group.

Acknowledgments

This publication is part of the Grant PID2020-114148RB-I00 funded by MCIN/AEI/10.13039/501100011033 and by ERDF A way of making Europe.

Conflict of interest

The authors declare that the research was conducted in the absence of any commercial or financial relationships that could be construed as a potential conflict of interest.

Author contributions

S.F., I.S., and M.B. wrote the manuscript and prepared the figures. V.C. and M.I. conceived the manuscript and the writing of the manuscript and approved its final content. Conceptualization, V.C. and M.I; writing original draft preparation, M.I.; reviewing and editing, V.C. and M.I.

References

Abrahamsen, G., Sundvold-Gjerstad, V., Habtamu, M., Bogen, B., & Spurkland, A. (2018). Polarity of CD4+ T cells towards the antigen presenting cell is regulated by the Lck adapter TSAd. *Scientific Reports*, *8*, 13319.

Alonso, R., Mazzeo, C., Rodriguez, M. C., Marsh, M., Fraile-Ramos, A., Calvo, V., et al. (2011). Diacylglycerol kinase alpha regulates the formation and polarisation of mature multivesicular bodies involved in the secretion of Fas ligand-containing exosomes in T lymphocytes. *Cell Death and Differentiation*, *18*, 1161–1173.

Bello-Gamboa, A., Izquierdo, J. M., Velasco, M., Moreno, S., Garrido, A., Meyers, L., et al. (2019). Imaging the human immunological synapse. *Journal of Visualized Experiments*, *154*, e60312.

Bello-Gamboa, A., Velasco, M., Moreno, S., Herranz, G., Ilie, R., Huetos, S., et al. (2020). Actin reorganization at the centrosomal area and the immune synapse regulates polarized secretory traffic of multivesicular bodies in T lymphocytes. *J Extracell Vesicles*, *9*, 1759926.

Ben-Shmuel, A., Sabag, B., Biber, G., & Barda-Saad, M. (2021). The role of the cytoskeleton in regulating the natural killer cell immune response in health and disease: From signaling dynamics to function. *Frontiers in Cell and Developmental Biology*, *9*, 609532.

Billadeau, D. D., Nolz, J. C., & Gomez, T. S. (2007). Regulation of T-cell activation by the cytoskeleton. *Nature Reviews. Immunology*, *7*, 131–143.

Blumenthal, D., & Burkhardt, J. K. (2020). Multiple actin networks coordinate mechanotransduction at the immunological synapse. *The Journal of Cell Biology*, *219*, e201911058.

Brown, A. C., Oddos, S., Dobbie, I. M., Alakoskela, J. M., Parton, R. M., Eissmann, P., et al. (2011). Remodelling of cortical actin where lytic granules dock at natural killer cell immune synapses revealed by super-resolution microscopy. *PLoS Biology*, *9*, e1001152.

Calabia-Linares, C., Robles-Valero, J., De La Fuente, H., Perez-Martinez, M., Martin-Cofreces, N., Alfonso-Perez, M., et al. (2011). Endosomal clathrin drives actin accumulation at the immunological synapse. *Journal of Cell Science*, *124*, 820–830.

Calvo, V., & Izquierdo, M. (2018). Imaging polarized secretory traffic at the immune synapse in living T lymphocytes. *Frontiers in Immunology*, *9*, 684.

Calvo, V., & Izquierdo, M. (2020). Inducible polarized secretion of exosomes in T and B lymphocytes. *International Journal of Molecular Sciences*, *21*, 2631.

Calvo, V., & Izquierdo, M. (2021). Role of actin cytoskeleton reorganization in polarized secretory traffic at the immunological synapse. *Frontiers in Cell and Developmental Biology*, *9*, 629097.

Carisey, A. F., Mace, E. M., Saeed, M. B., Davis, D. M., & Orange, J. S. (2018). Nanoscale dynamism of actin enables secretory function in Cytolytic cells. *Current Biology*, *28*(489–502), e489.

Chemin, K., Bohineust, A., Dogniaux, S., Tourret, M., Guegan, S., Miro, F., et al. (2012). Cytokine secretion by CD4+ T cells at the immunological synapse requires Cdc42-dependent local actin remodeling but not microtubule organizing center polarity. *Journal of Immunology*, *189*, 2159–2168.

Colon-Franco, J. M., Gomez, T. S., & Billadeau, D. D. (2011). Dynamic remodeling of the actin cytoskeleton by FMNL1gamma is required for structural maintenance of the Golgi complex. *Journal of Cell Science*, *124*, 3118–3126.

De La Roche, M., Asano, Y., & Griffiths, G. M. (2016). Origins of the cytolytic synapse. *Nature Reviews. Immunology*, *16*, 421–432.

Dogterom, M., & Koenderink, G. H. (2019). Actin–microtubule crosstalk in cell biology. *Nature Reviews Molecular Cell Biology*, *20*, 38–54.

Fooksman, D. R., Vardhana, S., Vasiliver-Shamis, G., Liese, J., Blair, D. A., Waite, J., et al. (2010). Functional anatomy of T cell activation and synapse formation. *Annual Review of Immunology*, *28*, 79–105.

Fritzsche, M., Fernandes, R. A., Chang, V. T., Colin-York, H., Clausen, M. P., Felce, J. H., et al. (2017). Cytoskeletal actin dynamics shape a ramifying actin network underpinning immunological synapse formation. *Science Advances*, *3*, e1603032.

Gomez, T. S., Kumar, K., Medeiros, R. B., Shimizu, Y., Leibson, P. J., Billadeau, D., et al. (2007). Formins regulate the actin-related protein 2/3 complex-independent polarization of the centrosome to the immunological synapse. *Immunity*, *26*, 177–190.

Griffiths, G. M., Tsun, A., & Stinchcombe, J. C. (2010). The immunological synapse: A focal point for endocytosis and exocytosis. *The Journal of Cell Biology*, *189*, 399–406.

Hammer, J. A., Wang, J. C., Saeed, M., & Pedrosa, A. T. (2018). Origin, organization, dynamics, and function of actin and Actomyosin networks at the T cell immunological synapse. *Annual Review of Immunology*, *37*, 201–224.

References

Herranz, G., Aguilera, P., Davila, S., Sanchez, A., Stancu, B., Gomez, J., et al. (2019). Protein kinase C delta regulates the depletion of actin at the immunological synapse required for polarized exosome secretion by T cells. *Frontiers in Immunology, 10*, 851.

Huppa, J. B., & Davis, M. M. (2003). T-cell-antigen recognition and the immunological synapse. *Nature Reviews. Immunology, 3*, 973–983.

Huse, M. (2012). Microtubule-organizing center polarity and the immunological synapse: Protein kinase C and beyond. *Frontiers in Immunology, 3*, 235.

Huse, M., Quann, E. J., & Davis, M. M. (2008). Shouts, whispers and the kiss of death: Directional secretion in T cells. *Nature Immunology, 9*, 1105–1111.

Ibanez-Vega, J., Del Valle Batalla, F., Saez, J. J., Soza, A., & Yuseff, M. I. (2019). Proteasome dependent actin remodeling facilitates antigen extraction at the immune synapse of B cells. *Frontiers in Immunology, 10*, 225.

Kumari, S., Curado, S., Mayya, V., & Dustin, M. L. (2014). T cell antigen receptor activation and actin cytoskeleton remodeling. *Biochimica et Biophysica Acta, 1838*, 546–556. https://doi.org/10.1016/j.bbamem.2013.1005.1004.

Kuokkanen, E., Sustar, V., & Mattila, P. K. (2015). Molecular control of B cell activation and immunological synapse formation. *Traffic, 16*, 311–326.

Lagrue, K., Carisey, A., Oszmiana, A., Kennedy, P. R., Williamson, D. J., Cartwright, A., et al. (2013). The central role of the cytoskeleton in mechanisms and functions of the NK cell immune synapse. *Immunological Reviews, 256*, 203–221.

Le Floc'h, A., & Huse, M. (2015). Molecular mechanisms and functional implications of polarized actin remodeling at the T cell immunological synapse. *Cellular and Molecular Life Sciences, 72*, 537–556.

Mazzeo, C., Calvo, V., Alonso, R., Mérida, I., & Izquierdo, M. (2016). Protein kinase D1/2 is involved in the maturation of multivesicular bodies and secretion of exosomes in T and B lymphocytes. *Cell Death and Differentiation, 23*, 99–109.

Monks, C. R. F., Freiberg, B. A., Kupfer, H., Sciaky, N., & Kupfer, A. (1998). Three-dimensional segregation of supramolecular activation clusters in T cells. *Nature, 395*, 82–86.

Montoya, M. C., Sancho, D., Bonello, G., Collette, Y., Langlet, C., He, H. T., et al. (2002). Role of ICAM-3 in the initial interaction of T lymphocytes and APCs. *Nature Immunology, 3*, 159–168.

Obino, D., Diaz, J., Saez, J. J., Ibanez-Vega, J., Saez, P. J., Alamo, M., et al. (2017). Vamp-7-dependent secretion at the immune synapse regulates antigen extraction and presentation in B-lymphocytes. *Molecular Biology of the Cell, 28*, 890–897.

Obino, D., Farina, F., Malbec, O., Saez, P. J., Maurin, M., Gaillard, J., et al. (2016). Actin nucleation at the centrosome controls lymphocyte polarity. *Nature Communications, 7*, 10969.

Peters, P. J., Borst, J., Oorschot, V., Fukuda, M., Krahenbuhl, O., Tschopp, J., et al. (1991). Cytotoxic T lymphocyte granules are secretory lysosomes, containing both perforin and granzymes. *The Journal of Experimental Medicine, 173*, 1099–1109.

Rak, G. D., Mace, E. M., Banerjee, P. P., Svitkina, T., & Orange, J. S. (2011). Natural killer cell lytic granule secretion occurs through a pervasive actin network at the immune synapse. *PLoS Biology, 9*, e1001151.

Ritter, A. T., Angus, K. L., & Griffiths, G. M. (2013). The role of the cytoskeleton at the immunological synapse. *Immunological Reviews, 256*, 107–117.

Ritter, A. T., Asano, Y., Stinchcombe, J. C., Dieckmann, N. M., Chen, B. C., Gawden-Bone, C., et al. (2015). Actin depletion initiates events leading to granule secretion at the immunological synapse. *Immunity, 42*, 864–876.

Ritter, A. T., Kapnick, S. M., Murugesan, S., Schwartzberg, P. L., Griffiths, G. M., & Lippincott-Schwartz, J. (2017). Cortical actin recovery at the immunological synapse leads to termination of lytic granule secretion in cytotoxic T lymphocytes. *Proceedings of the National Academy of Sciences of the United States of America*, *114*, E6585–E6594.

Saez, J. J., Diaz, J., Ibanez, J., Bozo, J. P., Cabrera Reyes, F., Alamo, M., et al. (2019). The exocyst controls lysosome secretion and antigen extraction at the immune synapse of B cells. *The Journal of Cell Biology*, *218*, 2247–2264.

Sanchez, E., Liu, X., & Huse, M. (2019). Actin clearance promotes polarized dynein accumulation at the immunological synapse. *PLoS One*, *14*, e0210377.

Stinchcombe, J. C., Majorovits, E., Bossi, G., Fuller, S., & Griffiths, G. M. (2006). Centrosome polarization delivers secretory granules to the immunological synapse. *Nature*, *443*, 462–465.

Ueda, H., Morphew, M. K., Mcintosh, J. R., & Davis, M. M. (2011). CD4+ T-cell synapses involve multiple distinct stages. *Proceedings of the National Academy of Sciences of the United States of America*, *108*, 17099–17104.

Yuseff, M. I., Lankar, D., & Lennon-Dumenil, A. M. (2009). Dynamics of membrane trafficking downstream of B and T cell receptor engagement: Impact on immune synapses. *Traffic*, *10*, 629–636.

Yuseff, M. I., Pierobon, P., Reversat, A., & Lennon-Dumenil, A. M. (2013). How B cells capture, process and present antigens: A crucial role for cell polarity. *Nature Reviews. Immunology*, *13*, 475–486.

CHAPTER 3

P815-based redirected degranulation assay to study human NK cell effector functions

Iñigo Terrén[a,†], Gabirel Astarloa-Pando[a,†], Ainhoa Amarilla-Irusta[a], and Francisco Borrego[a,b,*]

[a]Biocruces Bizkaia Health Research Institute, Immunopathology Group, Barakaldo, Spain
[b]Ikerbasque, Basque Foundation for Science, Bilbao, Spain
*Corresponding author: e-mail address: francisco.borregorabasco@osakidetza.eus

Chapter outline

1 Introduction..34
2 Prior to the assay...37
3 Redirected degranulation assay..37
4 Flow cytometry staining..38
5 Analyzing CD16-induced effector functions of NK cells..................................41
6 Concluding remarks..44
7 Notes..44
Acknowledgments..46
Conflict of interests...46
References..46

Abstract

Natural killer (NK) cells are part of the innate immune system, the classic cytotoxic population of innate lymphoid cells (ILCs). They can directly kill virus-infected or tumor cells through different mechanisms without prior sensitization using their lytic functions in response to different signals (target cell ligands and/or inflammatory cytokines) and secreting cytokines, such as interferon gamma (IFNγ) and tumor necrosis factor (TNF). NK cells use antibody-dependent cell-mediated cytotoxicity (ADCC) to recognize and kill cells expressing target

[†]These authors have contributed equally to this work.

antigens when they are antibody coated. Redirected cytotoxicity is a technique used to target cells that do not per se activate NK cells. Here, we use redirected degranulation, a surrogate technique that correlates with redirected lysis. The P815 cell line (mouse mastocytoma) express fragment crystallizable gamma receptor II (FcγRII) and therefore could bind the Fc portion of mouse IgG antibodies, which through their fragment antigen-binding (Fab) may recognize NK cells activating receptors leading to target cell lysis. This technique could be used to determine the inhibitory or activating capacity of different receptors or isoforms and in immunotherapy using T cell and NK cell activators.

1 Introduction

Natural killer (NK) cells are members of the innate immune system and have been classified as part of the heterogeneous family of the innate lymphoid cells (ILCs). They represent the classic cytotoxic population of ILCs: they can directly kill target cells through different mechanisms without prior sensitization (Guillerey, 2018; Krabbendam, Bernink, & Spits, 2021; Vivier et al., 2018) and in this way they control viral infections and cancers. On the one hand, they use their lytic functions in response to different signals (target cell ligands and/or inflammatory cytokines) and on the other hand, they secrete cytokines, such as interferon gamma (IFNγ) and tumor necrosis factor (TNF) (Mjösberg & Spits, 2016).

Regarding the lytic function, NK cells can release cytolytic granules to the immunological synapse causing the death of target cells. These granules contain perforin and granzymes: perforin forms pores in the plasma membrane enabling granzymes to enter into target cells, and granzymes induce target cell apoptosis (Lopez et al., 2013; Prager & Watzl, 2019). In order to regulate this process, NK cells express activating and inhibitory receptors whose integrated signals will determine their response. In humans, NK cell receptors include, among others, the inhibitory and activating forms of KIRs (killer Ig-like receptors), CD94/NKG2 receptors, NCRs (natural cytotoxicity receptors) including NKp30, NKp44 and NKp46, 2B4 receptor, NKp80, DNAM-1 and CD16 (Bottino, Biassoni, Millo, Moretta, & Moretta, 2000; Di Vito et al., 2019). Receptor–ligand interactions are of two types: (A) Recognition of autologous determinants, such as self-major histocompatibility complex class I (MHC-I) antigens by inhibitory KIRs and CD94/NKG2A receptor, will result in tolerance of NK cells toward self; (B) Detection of stress-induced self-molecules, which are normally expressed at very low levels and are upregulated in virus-infected and cancer cells, leading to NK cell activation and elimination of tumor and infected cells (Vivier, Ugolini, Blaise, Chabannon, & Brossay, 2012). Thus, there could be four different situations: (1) Tolerance: NK cells do not kill healthy cells expressing normal levels of self MHC-I molecules and few (or none) activating ligands; (2) Induced-self killing: NK cells recognize and kill transformed cells that express high amounts of activating ligands, even in the presence of normal levels of self MHC-I molecules; (3) Missing-self with killing: NK cells kill transformed cells that exhibit a reduced expression or mismatched expression of MHC-I molecules and high levels of activating ligands; (4) Missing-self without killing: when target cells have low (or none) amounts of activating ligands and the activating signal is low, NK cells do not kill transformed cells even when they have a reduced expression or mismatched expression of MHC-I molecules (Orrantia, Terrén, Astarloa-Pando, Zenarruzabeitia, & Borrego, 2021). Apart from that, NK cells use fragment crystallizable (Fc) receptors to bind the Fc region of IgG antibodies and release cytotoxic granules in order to kill target cells. Many tumors or virus-infected cells upregulate the expression of some antigens, which could be recognized by antibodies. NK cells express the activating receptor CD16a (FcγRIIIa) that recognize the Fc portion of IgG antibodies. In this way, they can kill antibody-coated target cells via antibody-dependent cell-mediated cytotoxicity (ADCC) (Lo Nigro et al., 2019). Furthermore, NK cell activation is also regulated by cytokines, such as interleukin (IL)-2, IL-12, IL-15 and IL-18, secreted by other cells (Zwirner & Domaica, 2010).

NK cells are able of secreting IFNγ, TNF and some chemokines such as C—C motif chemokine ligand 4 (CCL4). This response is also regulated by the integrated signaling cascades derived from activating and inhibitory receptors and by cytokine-induced signaling. CD107a, a membrane protein located in cytolytic granules, is used as a marker of NK cell function because the expression of this molecule correlates with TNF and IFNγ secretion and target cell lysis. IFNγ and TNF have been shown to be essential in viral and tumor eradication (Alter, Malenfant, & Altfeld, 2004; Balkwill, 2009; Ikeda, Old, & Schreiber, 2002; Jorgovanovic, Song, Wang, & Zhang, 2020).

Human NK cells can be divided in two major subsets, depending on the expression of CD56 and CD16: $CD56^{bright}CD16low/-$ ($CD56^{bright}$) and $CD56^{dim}CD16+$ ($CD56^{dim}$). These two subsets have important differences regarding their functionality (Cooper, Fehniger, & Caligiuri, 2001). $CD56^{bright}$ NK cells express low amounts of granzymes A and B and perforin, and low or undetectable amounts of CD16. They produce large amounts of immunomodulatory cytokines and chemokines in response to inflammatory cytokines but have low cytotoxicity unless they are activated via cytokines. On the other hand, $CD56^{dim}$ NK cells express high amounts of CD16, perforin and granzymes. They are more cytotoxic but produce lower amounts of cytokines unless they are stimulated by susceptible target cells (Orrantia et al., 2021).

As mentioned above, ADCC is a mechanism by which NK cells recognize and kill target cells expressing antigens when they are antibody-coated (Fig. 1A). Some cells do not express ligands that could be recognized by NK cell activating receptors and therefore are not lysed. One example could be the P815 cell line (mouse mastocytoma). Nevertheless, P815 cells have receptors that are able to recognize the Fc fragment of (mouse) IgG antibodies: they express FcγRII (Benhamou, Bonnerot, Fridman, & Daëron, 1990). Redirected cytotoxicity (also referred to as redirected lysis) is a technique, which causes cell-mediated lysis of target cells that under normal conditions would not be lysed (Fig. 1B). Here, we use redirected degranulation assay, which is correlated with target cell lysis. There are some receptors able of inducing an activating signal strong enough to activate T cells (CD3 with or without

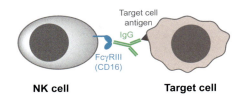

FIG. 1

Differences between antibody-dependent cell-mediated cytotoxicity (ADCC) (A) and redirected cytotoxicity (B). (A) In ADCC, NK cells are activated against target cells because they recognize target cell antigen-bound IgG antibodies (green) via FcγRIII (CD16). (B) In redirected cytotoxicity, P815 cell receptor FcγRII recognize the Fc fragment of anti-FcγRIII (CD16) IgG antibodies (purple), which binds to the FcγRIII expressed on NK cells inducing their activation.

co-stimulation through CD28) or NKT cells (CD3 or NKG2D alone) (Azuma, Cayabyab, Phillips, & Lanier, 1993; Kuylenstierna et al., 2011; Wilson et al., 1999). Regarding human NK cells, they can lyse P815 targets in the presence of mouse IgG anti-human CD16 antibodies that bind the FcγRII on P815 cells through the Fc fragment and CD16 on NK cells through the F(ab')2 fragments. In this way, NK cells will be activated and lyse P815 cells (Fig. 1B).

The P815-based redirected assays can be used for various purposes. One application is to determine the inhibitory capacity of different receptors. In people living with human immunodeficiency virus (PLHIV), it has been used redirected degranulation and cytokine production assays in order to measure the capacity of the CD300a inhibitory receptor to inhibit CD16-induced NK cell activation (Vitallé et al., 2019). Another option can be to determine the activating capacity of different receptors. In hepatitis C virus (HCV) cell models, it has been shown that the activating receptor NKp30 is downregulated in NK cells by direct interaction with HCV-infected cells and that this affected NKp30-induced cytotoxicity (Holder, Stapleton, Gallant, Russell, & Grant, 2013). What is more, redirected lysis assays can be used to differentiate between activating and inhibitory isoforms of the same receptor. In murine models, it has been shown that 2B4 isoforms 2B4L and 2B4S have activating and inhibitory properties respectively by using P815-based redirected assays (Schatzle et al., 1999).

Nowadays, some immunotherapeutic agents have a similar mechanism to redirected lysis in order to target cancer cells. These therapies are used against different cancer types and include T cell and NK cell activators. One example of these molecules is AFM13, a tetravalent bispecific anti-CD30/CD16A antibody used to induce CD16-dependent NK cell-mediated lysis of CD30 expressing Reed-Sternberg cells in patients with Hodgkin's lymphoma (Wu, Fu, Zhang, & Liu, 2015).

Therefore, P815-based redirected assays (lysis, degranulation and cytokine production) can be useful in different contexts. Here, we describe step-by-step how to perform a redirected degranulation and cytokine production assay and how to interpret the results.

2 Prior to the assay

2.1. Prepare culture medium for peripheral blood mononuclear cells (PBMCs) or purified NK cells by supplementing Rosewell Park Memorial Institute (RPMI) 1640 medium supplemented with GlutaMax (Gibco, ref. 72400054) with 10% heat-inactivated fetal bovine serum (FBS) (Note 7.1), 1% non-essential amino acids (Gibco, ref. 11140035), 1% Sodium Pyruvate (ThermoFisher, ref. 11360039), and 1% penicillin-streptomycin (Gibco, ref. 15140122). Sterilize culture media with 0.22 μm filters (Corning, ref. 431097) and store at 4 °C (Note 7.2). This culture media will be hereinafter referred to as cRPMI.

2.2. Prepare culture medium for P815 cells by supplementing cRPMI with 5 μg/mL Plasmocin Treatment (InvivoGen, ref. ant-mpt). Sterilize culture media with 0.22 μm filters and store at 4 °C (Note 7.2). This culture media will be hereinafter referred to as P815 medium.

2.3. One week before the assay, culture P815 cells in the incubator at 37 °C and 5% CO_2 in a T75 flask. Subculture the cells every 2–3 days by inoculating 50,000–100,000 cell/mL in 20–30 mL of P815 medium (plasmocin-containing cRPMI) to ensure a sufficient cell number for the redirected degranulation assay (Note 7.3).

3 Redirected degranulation assay

3.1. Prepare a single cell suspension of PBMCs or purified NK cells [described in (Terrén, Orrantia, Vitallé, Zenarruzabeitia, & Borrego, 2020)]. Redirected degranulation assay must be performed in sterile conditions.

3.2. Prepare a single cell suspension of P815 cells.

- **3.3.** Centrifuge PBMCs and P815 cells at 300 g for 5 min at room temperature, and discard the supernatant.
- **3.4.** Resuspend PBMCs and P815 cells in sterile phosphate buffered saline (PBS) without calcium or magnesium (Lonza, ref. BE17-516F) (Note 7.4) and centrifuge at 300 g for 5 min at room temperature. Discard the supernatant.
- **3.5.** Resuspend PBMCs and P815 cells in cRPMI and count viable cells in a Neubauer chamber by Trypan Blue (Sigma-Aldrich, ref. T8154) exclusion method (Notes 7.5 and 7.6).
- **3.6.** Prepare single cell suspensions of PBMCs and P815 cells with a final concentration of 5×10^6 cells/mL.
- **3.7.** Plate 0.5×10^6 PBMCs (100 μL) and 0.5×10^6 P815 cells (100 μL) in each well in a 96 well round (U) bottom plate (Note 7.7). Final volume in each well should be 200 μL.
- **3.8.** Add the stimuli and appropriate isotype controls to the corresponding wells. For example, PBMCs can be stimulated with 2.5 μg/mL of mouse anti-human CD16 (clone 3G8, BD Biosciences, ref. 555404) (Note 7.8), and 2.5 μg/mL of mouse IgG1, κ (clone MOPC-21, BioLegend, ref. 400101) can be used as isotype control (Note 7.9).
- **3.9.** Add 2 μL/well of PE-labeled anti-CD107a (clone REA792, Miltenyi Biotec, ref. 130-111-621). This step allows to measure degranulation by flow cytometry.
- **3.10.** Pulse-centrifuge (10–15 s) 96-well plate at 200 g at room temperature, protected from light.
- **3.11.** Incubate cells for 1 h at 37 °C and 5% CO_2 in the incubator.
- **3.12.** Add to each well 0.66 μL/mL of GolgiStop (BD Biosciences, ref. 554724) and 1 μL/mL of GolgiPlug (BD Biosciences, ref. 555029) (Note 7.10). GolgiStop (monensin) and GolgiPlug (brefeldin A) are protein transport inhibitors that stop vesicle trafficking. This step allows to measure cytokine and chemokine production by flow cytometry.
- **3.13.** Pulse-centrifuge (10–15 s) 96-well plate at 200 g at room temperature, protected from light.
- **3.14.** Incubate cells for 5 additional hours at 37 °C and 5% CO_2 in the incubator.
- **3.15.** Cover the plate with aluminum foil and store at 4 °C (in the refrigerator) until flow cytometry staining is performed (Note 7.11).

4 Flow cytometry staining

- **4.1.** Transfer cells from 96-well plate to flow cytometry tubes. From now on, it is not necessary to work in sterile conditions.
- **4.2.** Viability staining:
  - **4.2.1.** Add 2 mL PBS and centrifuge cells at 300 g for 5 min at 4 °C. Discard supernatant.
  - **4.2.2.** Resuspend pellet in 1 mL PBS and add 1 μL of reconstituted LIVE/DEAD Fixable Near-IR (Invitrogen, ref. L34976) (Note 7.12).

- **4.2.3.** Incubate cells for 30 min at 4 °C, protected from light.
- **4.2.4.** Add 2 mL PBS supplemented with 2.5% bovine serum albumin (BSA) (Note 7.13) (Sigma-Aldrich, ref. 82100-M) and centrifuge cells at 300 g for 5 min at 4 °C. Discard supernatant and resuspend pellet.

4.3. Extracellular staining:
- **4.3.1.** If viability staining (step 4.2) has been skipped, wash cells with PBS containing 2.5% BSA as in step 4.2.4 before doing the extracellular staining.
- **4.3.2.** Incubate cells with the appropriate concentration of fluorochrome-labeled antibodies for 30 min at 4 °C, protected from light (Notes 7.12 and 7.14). The following mouse anti-human antibodies have been used in this protocol: BV421-labeled anti-CD56 (clone NCAM16.2, BD Biosciences, ref. 562751), BV510-labeled anti-CD3 (clone UCHT1, BD Biosciences, ref. 563109), BV510-labeled anti-CD14 (clone MφP9, BD Biosciences, ref. 563079), and BV510-labeled anti-CD19 (clone SJ25C1, BD Biosciences, ref. 562947) (Note 7.15). Also, for this protocol, 50 μL of Brilliant Stain Buffer (BD Biosciences, ref. 563794) have been added to each tube.
- **4.3.3.** Add 2 mL of PBS containing 2.5% BSA and centrifuge cells at 300 g for 5 min at 4 °C. Discard supernatant and resuspend pellet. Repeat this step.

4.4. Fixation and permeabilization:
- **4.4.1.** Resuspend cells in 250 μL of Fixation/Permeabilization Solution (BD Biosciences, ref. 554722) and vortex.
- **4.4.2.** Incubate cells for 20 min at 4 °C, protected from light.
- **4.4.3.** Add 2 mL of PermWash Buffer 1× (BD Biosciences, ref. 554723) (Note 7.16) and centrifuge cells at 300 g for 5 min at 4 °C. Discard supernatant and resuspend pellet.

4.5. Intracellular staining:
- **4.5.1.** Incubate cells with the appropriate concentration of fluorochrome-labeled antibodies for 30 min at 4 °C, protected from light (Notes 7.12 and 7.14). The following mouse anti-human antibodies has been used in this protocol: FITC-labeled anti-CCL4 (clone D21-1351, BD Biosciences, ref. 560565), PerCP-Cy5.5-labeled anti-IFNγ (clone B27, BD Biosciences, ref. 560704), and APC-labeled anti-TNF (clone Mab11, BioLegend, ref. 502912).
- **4.5.2.** Add 2 mL PBS containing 2.5% BSA and centrifuge cells at 300 g for 5 min at 4 °C. Discard supernatant and resuspend pellet. Repeat this step.

4.6. Resuspend cells in 350 μL PBS and acquire in a flow cytometer. We have run the samples in a LSRFortessa X-20 (Table 1) and acquired a minimum of 7000 NK cells per sample (Note 7.17).

Table 1 Antibodies and fluorescent dyes used for flow cytometry analysis

Laser	Filter	Fluorochrome	Marker	Manufacturer	Clone	Dilution
405 nm	450/50	BV421	CD56	BD Biosciences	NCAM16.2	1:20
	525/50	BV510	CD3	BD Biosciences	UCHT1	1:33
	525/50	BV510	CD14	BD Biosciences	MφP9	1:33
	525/50	BV510	CD19	BD Biosciences	SJ25C1	1:33
488 nm	530/30	FITC	CCL4	BD Biosciences	D21–1351	1:5
	575/25	PE	CD107a	Miltenyi Biotec	REA792	1:100
	710/50	PerCP-Cy5.5	IFNγ	BD Biosciences	B27	1:20
635 nm	670/30	APC	TNF	Biolegend	Mab11	1:33
	780/60	Near-IR	LIVE/DEAD	Invitrogen		1:1000

5 Analyzing CD16-induced effector functions of NK cells

In this section, we will analyze the results of a redirected degranulation assay performed with human PBMCs stimulated through CD16, as indicated in the previous sections. In addition, we have pre-activated PBMCs by stimulating them with 10 ng/mL of recombinant human interleukin (rhIL)-12 (Miltenyi Biotec, ref. 130-096-705), 100 ng/mL rhIL-15 (Miltenyi Biotec, ref. 130-095-764) and 50 ng/mL rhIL-18 (MBL International, ref. B001-5) for 16 h prior to performing the redirected degranulation assay, following previously published experimental settings (Terrén et al., 2018, 2021). In this way, we will be able to study CD16-induced degranulation (CD107a) and cytokine (IFNγ and TNF) and chemokine (CCL4) production of resting and pre-activated human NK cells.

We have performed data analysis with FlowJo v.10.7.1 software, although other analysis softwares can be used as well. We started by exporting data from the flow cytometer in FCS 3.0 format, and importing to the analysis software. First, we recommend checking the time parameter to exclude possible artifacts that can happen during the acquisition (Fig. 2A). Next, lymphocytes are gated attending to forward and side scatter areas (FSC-A and SSC-A, respectively). Single cells are then gated attending to FSC-A and FSC-H (height), and dead cells are excluded attending to LIVE/DEAD Near-IR. Finally, NK cells are gated based on the expression of CD3, CD14, CD19, and CD56 (Fig. 2B). Of note, we used an exclusion channel in which CD3, CD14 and CD19 markers are included because NK cells do not express any of them. This exclusion channel limited our analysis to study only NK cells. If there is an interest in studying other cell subsets (e.g., T cells), then anti-CD3, anti-CD14 and anti-CD19 should be labeled with different fluorochromes and the subsets of interest should be gated following an appropriate strategy (Cossarizza et al., 2021).

Once NK cells are gated, we have determined the percentage of NK cells (in the presence of P815 targets) that are positive for each function (Fig. 3). We have defined positive gates for each studied parameter attending to the negative control, i.e., resting NK cells stimulated with isotype control. Note that pre-activated NK cells stimulated with isotype control showed higher degranulation and cytokine production than resting NK cells. Therefore, defining a positive gate for each function in pre-activated NK cells would be challenging without the resting condition. In addition to determining the percentage of cells that are positive for each function, the median fluorescence intensity (MFI) can be calculated. This parameter could be useful to estimate the differences in cytokine and chemokine production per cell. For instance, pre-activated NK cells stimulated with isotype control and anti-CD16 showed a similar percentage of IFNγ+ cells (90.6% and 94.6%, respectively) (Fig. 3). However, the difference in IFNγ production is more evident attending to the MFI values. Pre-activated NK cells that have been further stimulated through CD16 have a higher MFI of IFNγ (within IFNγ+ NK cells) than pre-activated NK cells stimulated with isotype control (MFI of 13,419 and 8873, respectively) (Table 2). It must be emphasized that some authors analyze the MFI of each parameter within all NK cells, while

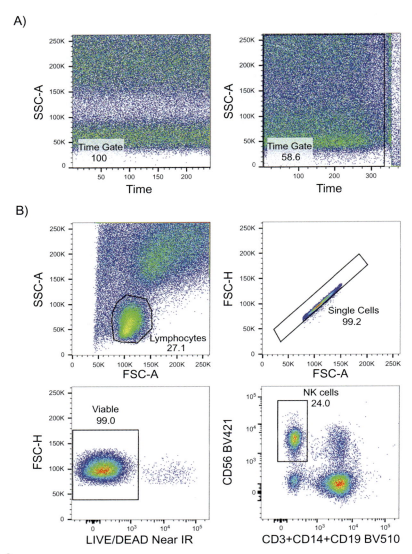

FIG. 2

Gating strategy of human NK cells. (A) Pseudocolor plots of flow cytometry data with (right) and without (left) any kind of problem during the acquisition. A "Time Gate" is used to select data without glitches or difficulties attending to Time and Side Scatter Area (SSC-A) parameters. (B) Pseudocolor plots showing the gating strategy. First, lymphocytes are gated attending to forward scatter area (FSC-A) and SSC-A. Single cells are then gated attending to FSC-A and FSC-H (height), and dead cells are excluded attending to the expression of LIVE/DEAD dye. NK cells are gated within viable lymphocytes based on the expression of CD3, CD14, CD19, and CD56.

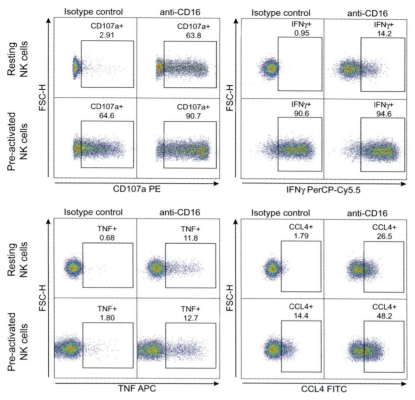

FIG. 3

CD16-induced effector functions of resting and pre-activated human NK cells. Pseudocolor plots showing the percentage of resting and pre-activated NK cells that degranulate (CD107a+), and produce cytokines (IFNγ+ and TNF+) and chemokines (CCL4+). Pre-activated NK cells were stimulated for 16h with IL-12, IL-15 and IL-18 (10, 100 and 50ng/mL, respectively) while resting NK cells were cultured without cytokines for the same time. Then, redirected degranulation and cytokine production assay was performed by co-culturing resting and pre-activated NK cells with P815 cells at an E:T ratio of 1:1 for 6h in the presence of 2.5μg/mL of mouse anti-CD16 (clone 3G8) or isotype control (clone MOPC-21).

Table 2 Median fluorescence intensity of different parameters in CD16-stimulated NK cells

	Resting NK cells		Pre-activated NK cells	
	Isotype control	Anti-CD16	Isotype control	Anti-CD16
MFI CD107a (CD107a+ NK cells)	897	4791	1950	10,821
MFI IFNγ (IFNγ+ NK cells)	1636	2676	8873	13,419
MFI TNF (TNF+ NK cells)	786	1788	1630	3532
MFI CCL4 (CCL4+ NK cells)	2387	3004	2882	3583

Median fluorescence intensity (MFI) of CD107a, IFNγ, TNF, and CCL4, within the NK cell subsets that are positive for each parameter from the experiment shown in Fig. 3. The determinations were done in resting and pre-activated NK cells.

others determine the MFI of a specific parameter within the NK cell subset that is positive for that parameter. The former strategy could be useful when positive and negative regions of a determined parameter are poorly differentiated, but the latter estimates more precisely cytokine and chemokine production per cell because negative cells (i.e., cells that are not positive for the studied function) are excluded of the analysis. The same experiment could led to different results depending on the analysis strategy that is followed. It should be also noted that, in addition to the median, other measurements of central tendency could be used to analyze fluorescence intensity, such as the arithmetic mean and geometric mean. Arithmetic mean is appropriate when data showed a normal distribution, but it should be considered that fluorescence intensity increases logarithmically. Geometric mean is appropriate when data showed a log-normal distribution, but it is susceptible to outliers. Unlike these measurements, median is neither affected by data distribution nor strongly influenced by outliers. Therefore, the consensus is that the median is the most reliable measurement, followed by the geometric mean when the outliers are of interest (Cossarizza et al., 2019). In any case, the chosen measurement must be always described to avoid misunderstandings when the term "MFI" is used.

6 Concluding remarks

NK cells use ADCC to recognize and kill cells expressing target antigens when they are antibody-coated. Redirected cytotoxicity is a technique used to target cells that do not spontaneously activate NK cells. Redirected degranulation (and cytokine production) is a surrogate of redirected lysis. The P815 mouse mastocytoma cell line can bind mouse IgG antibodies, which recognize NK cells activating receptors causing target cell lysis. This mechanism can be used to determine the inhibitory or activating capacity of different receptors or isoforms and in cancer immunotherapy.

7 Notes

7.1. Heat-inactivate FBS by incubating FBS at 56°C for 30 min in a water bath. Shake regularly FBS bottle during the incubation. Let the FBS cool down and aliquot heat-inactivated FBS in 50 mL conical tubes in sterile conditions. Freeze the aliquots until they are needed.

7.2. Store media at 4°C for no longer than a month.

7.3. P815 cell line should be routinely tested for mycoplasma contamination with appropriate tools, such as Venor GeM Classic detection kit (Minerva Biolabs, ref. 11-1100). Medium renewal should be done every 2–3 days. We suggest avoiding a cell density higher than $1–1.5 \times 10^6$ cell/mL.

7.4. PBS can be replaced by sterile RPMI 1640 or cRPMI in this step.

7.5. Trypan Blue stains cells with compromized cell membranes (i.e., dead cells). Count only viable (not Trypan Blue-stained) cells.

7.6. Redirected degranulation assay is sensitive to changes in cell numbers. Distinct effector to target cell (E:T) ratios would change the outcome of the assay. Thus, to obtain reliable and consistent results, it is crucial to count cells with precision.

7.7. This protocol with an E:T ratio of 1:1 has been optimized to co-culture PBMCs and P815 cells. If isolated NK cells will be used instead of PBMCs, different E:T ratios should be checked.

7.8. Usually, a concentration of up to 5 µg/mL is used in redirected assays, but we recommend to titrate the antibodies. Also, this protocol is designed to study CD16-induced degranulation and cytokine production in NK cells, but other mouse anti-human antibodies could be used instead of, or in combination with, anti-CD16 antibodies. This includes antibodies directed against both activating (e.g., anti-NKG2C, anti-NKG2D, anti-NKp30, or anti-NKp46) or inhibitory (e.g., anti-NKG2A or anti-CD300a) receptors.

7.9. We recommend adding an additional well with effector (PBMCs or NK cells) and target cells that could be used as an unstained control in the flow cytometry analysis.

7.10. GolgiStop and GolgiPlug should be stored at 4 °C. We recommend bringing them to room temperature during the step 3.11.

7.11. Plate can be stored overnight at 4 °C. This protocol is designed to be done over 2 days. During the first day, PBMCs or isolated NK cells are obtained and redirected degranulation assay is performed. Next day, samples are stained, acquired in the flow cytometer and analyzed. However, this protocol can be modified to perform the whole experiment in the same day.

7.12. Vortex the tubes and ensure to mix well the cells and dyes or antibodies before incubating them.

7.13. BSA can be replaced by human AB Serum in this step.

7.14. Antibodies should be previously titrated. Information about the antibody dilutions used in this experiment can be found in Table 1.

7.15. Note that anti-CD3, anti-CD14, and anti-CD19 are BV510-labeled. The reason is that the proposed experiment is designed to identify NK cells by using an exclusion channel with antibodies specific for CD3, CD14, and CD19 (NK cells do not express any of these markers). However, other extracellular antibodies can be used to study other cell subsets.

7.16. PermWash 1X is obtained by diluting 1:10 PermWash 10X with distillated water. We recommend doing this dilution during the step 4.4.2.

7.17. We strongly recommend performing a compensation with single-stained controls prior to data acquisition. This can be done days or weeks before the redirected degranulation assay.

Acknowledgments
This study was supported by the following grants: Fundación AECC-Spanish Association Against Cancer Foundation (PROYE16074BORR) and Health Department, Basque Government (2021333006). Iñigo Terrén is recipient of a predoctoral contract funded by the Department of Education, Basque Government (PRE_2021_2_0215). Gabirel Astarloa-Pando is recipient of a predoctoral contract funded by AECC-Spanish Association Against Cancer (PRDVZ21440ASTA). Gabirel Astarloa-Pando and Ainhoa Amarilla-Irusta are recipients of grants from Jesús de Gangoiti Barrera Foundation (FJGB21/001 and FJBG21/005). Francisco Borrego is an Ikerbasque Research Professor, Ikerbasque, Basque Foundation for Science.

Conflict of interests
The authors declare that the research was conducted in the absence of any commercial or financial relationships that could be construed as a potential conflict of interest.

References
Alter, G., Malenfant, J. M., & Altfeld, M. (2004). CD107a as a functional marker for the identification of natural killer cell activity. *Journal of Immunological Methods*, *294*(1–2), 15–22. https://doi.org/10.1016/j.jim.2004.08.008.

Azuma, M., Cayabyab, M., Phillips, J. H., & Lanier, L. L. (1993). Requirements for CD28 co-stimulation of T cell-mediated cytotoxicity. *Journal of Immunology*, *150*(6), 2091–2201.

Balkwill, F. (2009). Tumour necrosis factor and cancer. *Nature Reviews Cancer*, *9*(5), 361–371. https://doi.org/10.1038/nrc2628.

Benhamou, M., Bonnerot, C., Fridman, W. H., & Daëron, M. (1990). Molecular heterogeneity of murine mast cell fc gamma receptors. *Journal of Immunology*, *144*(8), 3071–3077.

Bottino, C., Biassoni, R., Millo, R., Moretta, L., & Moretta, A. (2000). The human natural cytotoxicity receptors (NCR) that induce HLA class I-independent NK cell triggering. *Human Immunology*, *61*(1), 1–6. https://doi.org/10.1016/S0198-8859(99)00162-7.

Cooper, M. A., Fehniger, T. A., & Caligiuri, M. A. (2001). The biology of human natural killer-cell subsets. *Trends in Immunology*, *22*(11), 633–640. https://doi.org/10.1016/S1471-4906(01)02060-9.

Cossarizza, A., Chang, H., Radbruch, A., Abrignani, S., Addo, R., Akdis, M., et al. (2021). Guidelines for the use of flow cytometry and cell sorting in immunological studies (third edition). *European Journal of Immunology*, *51*(12), 2708–3145. https://doi.org/10.1002/eji.202170126.

Cossarizza, A., Chang, H., Radbruch, A., Acs, A., Adam, D., Adam-Klages, S., et al. (2019). Guidelines for the use of flow cytometry and cell sorting in immunological studies (second edition). *European Journal of Immunology*, *49*(10), 1457–1973. https://doi.org/10.1002/eji.201970107.

References

Di Vito, C., Mikulak, J., Zaghi, E., Pesce, S., Marcenaro, E., & Mavilio, D. (2019). NK cells to cure cancer. In *Seminars in Immunology, March, 0–1*. https://doi.org/10.1016/j.smim.2019.03.004.

Guillerey, C. (2018). Roles of cytotoxic and helper innate lymphoid cells in cancer. In *Mammalian genome (vol. 29, issues 11–12)* (pp. 777–789). US: Springer. https://doi.org/10.1007/s00335-018-9781-4.

Holder, K. A., Stapleton, S. N., Gallant, M. E., Russell, R. S., & Grant, M. D. (2013). Hepatitis C virus–infected cells downregulate NKp30 and inhibit ex vivo NK cell functions. *Journal of Immunology, 191*(6), 3308–3318. https://doi.org/10.4049/jimmunol.1300164.

Ikeda, H., Old, L. J., & Schreiber, R. D. (2002). The roles of IFNγ in protection against tumor development and cancer immunoediting. *Cytokine and Growth Factor Reviews, 13*(2), 95–109. https://doi.org/10.1016/S1359-6101(01)00038-7.

Jorgovanovic, D., Song, M., Wang, L., & Zhang, Y. (2020). Roles of IFN-γ in tumor progression and regression: A review. *Biomarker Research, 8*(1), 1–16. https://doi.org/10.1186/s40364-020-00228-x.

Krabbendam, L., Bernink, J. H., & Spits, H. (2021). Innate lymphoid cells: From helper to killer. *Current Opinion in Immunology, 68*, 28–33. https://doi.org/10.1016/j.coi.2020.08.007.

Kuylenstierna, C., Björkström, N. K., Andersson, S. K., Sahlström, P., Bosnjak, L., Paquin-Proulx, D., et al. (2011). NKG2D performs two functions in invariant NKT cells: Direct TCR-independent activation of NK-like cytolysis and co-stimulation of activation by CD1d. *European Journal of Immunology, 41*(7), 1913–1923. https://doi.org/10.1002/eji.200940278.

Lo Nigro, C., Macagno, M., Sangiolo, D., Bertolaccini, L., Aglietta, M., & Merlano, M. C. (2019). NK-mediated antibody-dependent cell-mediated cytotoxicity in solid tumors: Biological evidence and clinical perspectives. *Annals of Translational Medicine, 7*(5), 105. https://doi.org/10.21037/atm.2019.01.42.

Lopez, J. A., Susanto, O., Jenkins, M. R., Lukoyanova, N., Sutton, V. R., Law, R. H. P., et al. (2013). Perforin forms transient pores on the target cell plasma membrane to facilitate rapid access of granzymes during killer cell attack. *Blood, 121*(14), 2659–2668. https://doi.org/10.1182/blood-2012-07-446146.

Mjösberg, J., & Spits, H. (2016). Human innate lymphoid cells. *Journal of Allergy and Clinical Immunology, 138*(5), 1265–1276. https://doi.org/10.1016/j.jaci.2016.09.009.

Orrantia, A., Terrén, I., Astarloa-Pando, G., Zenarruzabeitia, O., & Borrego, F. (2021). Human NK cells in autologous hematopoietic stem cell transplantation for cancer treatment. *Cancers, 13*(7), 1589. https://doi.org/10.3390/cancers13071589.

Prager, I., & Watzl, C. (2019). Mechanisms of natural killer cell-mediated cellular cytotoxicity. *Journal of Leukocyte Biology, 105*(6), 1319–1329. https://doi.org/10.1002/JLB.MR0718-269R.

Schatzle, J. D., Sheu, S., Stepp, S. E., Mathew, P. A., Bennett, M., & Kumar, V. (1999). Characterization of inhibitory and stimulatory forms of the murine natural killer cell receptor 2B4. *Proceedings of the National Academy of Sciences of the United States of America, 96*(7), 3870–3875. https://doi.org/10.1073/pnas.96.7.3870.

Terrén, I., Mikelez, I., Odriozola, I., Gredilla, A., González, J., Orrantia, A., et al. (2018). Implication of Interleukin-12/15/18 and Ruxolitinib in the phenotype, proliferation, and polyfunctionality of human cytokine-preactivated natural killer cells. *Frontiers in Immunology, 9*, 737. https://doi.org/10.3389/fimmu.2018.00737.

Terrén, I., Orrantia, A., Mosteiro, A., Vitallé, J., Zenarruzabeitia, O., & Borrego, F. (2021). Metabolic changes of Interleukin-12/15/18-stimulated human NK cells. *Scientific Reports*, *11*(1), 6472. https://doi.org/10.1038/s41598-021-85960-6.

Terrén, I., Orrantia, A., Vitallé, J., Zenarruzabeitia, O., & Borrego, F. (2020). CFSE dilution to study human T and NK cell proliferation in vitro. In. *Methods in Enzymology*, *631*, 239–255. https://doi.org/10.1016/bs.mie.2019.05.020.

Vitallé, J., Terrén, I., Orrantia, A., Pérez-Garay, R., Vidal, F., Iribarren, J. A., et al. (2019). CD300a inhibits CD16-mediated NK cell effector functions in HIV-1-infected patients. *Cellular & Molecular Immunology*, *16*(12), 940–942. https://doi.org/10.1038/s41423-019-0275-4.

Vivier, E., Artis, D., Colonna, M., Diefenbach, A., Di Santo, J. P., Eberl, G., et al. (2018). Innate lymphoid cells: 10 years on. *Cell*, *174*(5), 1054–1066. https://doi.org/10.1016/j.cell.2018.07.017.

Vivier, E., Ugolini, S., Blaise, D., Chabannon, C., & Brossay, L. (2012). Targeting natural killer cells and natural killer T cells in cancer. *Nature Reviews Immunology*, *12*(4), 239–252. https://doi.org/10.1038/nri3174.

Wilson, J. L., Charo, J., Martín-Fontecha, A., Dellabona, P., Casorati, G., Chambers, B. J., et al. (1999). NK cell triggering by the human costimulatory molecules CD80 and CD86. *Journal of Immunology (Baltimore, Md.: 1950)*, *163*(8), 4207–4212.

Wu, J., Fu, J., Zhang, M., & Liu, D. (2015). AFM13: A first-in-class tetravalent bispecific anti-CD30/CD16A antibody for NK cell-mediated immunotherapy. *Journal of Hematology & Oncology*, *8*(1), 28–31. https://doi.org/10.1186/s13045-015-0188-3.

Zwirner, N. W., & Domaica, C. I. (2010). Cytokine regulation of natural killer cell effector functions. *BioFactors*, *36*(4), 274–288. https://doi.org/10.1002/biof.107.

CHAPTER 4

Cytotoxic and chemotactic dynamics of NK cells quantified by live-cell imaging

Yanting Zhu and Jue Shi*
Department of Physics and Department of Biology, Center for Quantitative Systems Biology, Hong Kong Baptist University, Hong Kong, China
Corresponding author: e-mail address: jshi@hkbu.edu.hk

Chapter outline

1 Introduction..50
2 Materials..51
   2.1 Common disposables...51
   2.2 Cells and reagents..52
   2.3 Microscope system..52
3 Methods...53
   3.1 Cell maintenance..53
   3.2 NK cell staining with the LysoBrite red dye..........................53
   3.3 Generation of stable fluorescent reporter cancer cell lines expressing the Granzyme B-FRET reporter and/or IMS-RP (mitochondria reporter)............54
   3.4 Transient expression of the Caspase 8-FRET fluorescent reporter in cancer cells..55
   3.5 Preparation of NK-cancer cell co-culture for the imaging experiments........56
   3.6 Phase-contrast and fluorescence live-cell imaging...................56
   3.7 Quantification of the NK-cancer cell interaction dynamics based on the imaging data..57
4 Notes..62
Acknowledgments...63
References..63

CHAPTER 4 Cytotoxic and chemotactic dynamics of NK cells

Abstract

Natural Killer (NK) cells detect and eliminate virus-infected cells and cancer cells, and are crucial players of the human immune defense system. Although the relevant molecular machineries involved in NK cell activation and NK-target cell interactions are largely known, how their collective signaling modulates the dynamic behaviors of NK cells, e.g., motility and cytotoxicity, and the rate-limiting kinetics involved are still in need of comprehensive investigations. In traditional bulk killing assays, heterogeneity and kinetic details of individual NK-target cell interactions are masked, seriously limiting analysis of the underlying dynamic mechanisms. Here we present detailed protocols of a number of live-cell imaging assays using fluorescent protein reporters and/or a live-cell dye that enable the acquisition of quantitative kinetic data at the single cell level for elucidating the mechanism underlying the interaction dynamics of primary human NK cells and epithelial cancer cells. Moreover, we discuss how the imaging data can be analyzed either alone or in combination to quantify and determine the key dynamic steps/intermediates involved in specific NK cell activity, e.g., NK cell cytotoxic modes and their associated kinetics, and NK cell motility toward different cancer targets. These live-cell imaging assays can be easily adapted to analyze the rate-limiting kinetics and heterogeneity of other cell-cell interaction dynamics, e.g., in T cell function.

1 Introduction

Natural Killer (NK) cells are cytotoxic lymphocytes analogous to cytotoxic T cells (CTLs), but they belong to the innate immune system, as they are capable of eliciting rapid immune response in the absence of antibodies or antigen matching with MHCI molecules (Bryceson, March, Ljunggren, & Long, 2006; Caligiuri, 2008). Such alloreactive characteristic confers NK cells attractive therapeutic potential for treating cancer and various chronic diseases, where immunomodulation plays an important role (Chiossone, Dumas, Vienne, & Vivier, 2018). The process of NK cell activation against an abnormal cell target is regulated in a sequence of dynamic steps, including target-directed NK cell chemotaxis, target recognition at the NK-target cell immunological synapse, NK cell killing of the abnormal targets and/or cytokine/chemokine secretion by the activated NK cells that generate and enhance systemic inflammatory signaling. Among the different dynamic steps, mechanisms underlying target recognition and NK cell killing of the abnormal targets are the most well studied (Bryceson et al., 2006; Lanier, 2005; Prager & Watzl, 2019).

NK cells are known to distinguish abnormal cells via signaling at the NK-target cell immunological synapse. Various inhibitory and activating NK ligands expressed on the cell targets bind to their cognate receptors on NK cells, and the resulting integrated signal determines whether NK cell activity is inhibited or activated (Bryceson et al., 2006; Lanier, 2005). The major group of inhibitory receptors expressed by human NK cells are Killer cell Immunoglobulin-like Receptors (KIRs) that bind to MHCI molecules. Compared to the small number of inhibitory receptors with common inhibitory ligands, there are a much wider spectrum of NK cell activating receptors, which probably evolved to respond to diverse aberrant ligands expressed by different types of virus-infected cells and transformed cells. For instance, tumor cells are found to activate antitumor activity of NK cells by upregulating a number of activating NK ligands, such as MICA (MHC class I-related gene A), MICB (MHC class I-related gene B) and ULBP (UL16-binding protein), as well as downregulating the MHCI molecules (Champsaur & Lanier, 2010; Drake, Jaffee, & Pardoll, 2006). Subsequent to recognizing an abnormal cell target, NK cells can elicit direct target cell killing through multiple cytotoxic signaling pathways, including lytic granule-mediated apoptosis, death ligand-mediated apoptosis, and the more recently discovered, pyroptosis and necroptosis (Prager & Watzl, 2019; Zhou et al., 2020; Zhu, Xie, & Shi, 2021). In addition to triggering direct cytotoxicity, another major effector function of NK cell is to secret cytokines and chemokines that activate other immune cell types and orchestrate the complex immune response at the system level (Lee, Miyagi, & Biron, 2007; Vivier, Tomasello, Baratin, Walzer, & Ugolini, 2008). Interferon gamma (IFN-γ) is a particularly important cytokine secreted by NK cells upon inflammatory response.

Although the molecular machineries discussed above are largely known, how their collected dynamics regulate fast yet highly selective target cell killing by NK cells in the complex environment of tissues is still poorly understood. Detailed analysis and elucidation of the rate-limiting kinetics underlying differential NK cell activities and the heterogeneity involved are key to advance our understanding of NK cell control as well as the target selectivity of NK cell function, providing a more informed basis to develop NK cell-based immunotherapy. Live-cell imaging is a powerful and versatile tool to decipher real-time interaction dynamics between different cell types at the single cell level. Quantitative single cell microscopy studies both by us and others have revealed intriguing new dynamic features of NK cell motility and novel cytotoxic mechanisms (Bhat & Watzl, 2007; Choi & Mitchison, 2013; Lee & Mace, 2017; Olofsson et al., 2014; Zhu et al., 2021; Zhu, Huang, & Shi, 2016). Below we detailed a few live-cell imaging assays based on phase-contrast and/or fluorescence imaging to measure NK cell chemotaxis toward cancer targets and the cytotoxic dynamics of NK cells.

2 Materials

2.1 Common disposables

1. 6-well plates, 12-well plates and 10-cm dishes for cell culture
2. 96-well imaging plates for live-cell imaging (μ-plate, ibidi, Germany)
3. Serological pipettes (5 mL and 10 mL)
4. Micropipettes and tips
5. 15 mL and 50 mL conical-bottom centrifuge tubes
6. 1.5 mL microcentrifuge tubes
7. Disposable cell counting slides and Neubauer hemocytometer
8. 0.45 μm sterile syringe filters

2.2 Cells and reagents
1. Primary human NK cells (isolated from fresh human PBMCs using the EasySep Human NK Cell Enrichment Kit (STEMCELL Technologies), according to manufacturer's protocol)
2. Cell culture media and medium supplements, including heat-inactivated Fetal Calf Serum, penicillin and streptomycin (Gibco, Thermo Fisher Scientific)
3. Phenol-red free, CO_2-independent medium (Gibco, Thermo Fisher Scientific)
4. Recombinant human Interleukine 2 (IL-2, PeproTech)
5. LysoBrite red dye (AAT Bioquest) for staining NK cell lytic granules
6. Granzyme B-FRET live-cell reporter construct: this retroviral report construct consists of a cyan (CFP, donor) and yellow (YFP, receptor) fluorescent protein linked by a peptide substrate specific to granzyme B, i.e., VGPDFGR (Choi & Mitchison, 2013). Upon lytic granule transfer and release of granzyme B from NK cells into the target cells, granzyme B cleaves the peptide linker of this FRET reporter expressed by the target cells, and the energy transfer from CFP to YFP is then lost, resulting in a decrease of YFP fluorescence and increase of CFP fluorescence. Therefore, NK cell cytotoxic activity mediated by granzyme B proteolytic activity can be specifically measured in real time by imaging this FRET live-cell reporter.
7. Caspase 8-FRET live-cell reporter construct: this report construct consists of a cyan (donor) and yellow (acceptor) fluorescent protein linked by a peptide substrate specific to caspase 8, i.e., GLRSGGIETDGGIETDGGSGST (Albeck et al., 2008). Upon death ligand-mediated extrinsic apoptosis triggered by NK cells, the caspase 8-FRET signal in the target cells is lost, allowing the measurement of the rate-limiting kinetics underlying this specific NK cell killing mode.
8. Mitochondria live-cell reporter construct, IMS-RP: this retroviral reporter construct encodes a monomeric red fluorescent protein targeted to the inter-membrane space of mitochondria by fusion to the leader peptide of SMAC (Albeck et al., 2008). Upon mitochondria outer membrane permeabilization (MOMP), the committed step of apoptosis, the reporter diffuses from mitochondria to the cytoplasm, resulting in an abrupt transition from punctate to diffused localization of the fluorescence signal. Therefore, the onset of apoptotic target cell death induced by the NK cell cytotoxic activity can be monitored and quantified by imaging this IMS-RP reporter.
9. X-tremeGene HP DNA transfection reagent (Roche)
10. PEG-it Virus Precipitation Solution (SBI, System Biosciences)
11. Polybrene (Millipore)

2.3 Microscope system
Nikon Eclipse Ti2 inverted microscope equipped with: a humidified incubator maintained at 37 °C for live-cell culture; phase annulus plates for phase-contrast imaging; a 20× phase-contrast objective (NA = 0.45); a 20× plan Apo objective with high

NA (NA = 0.75) for fluorescence imaging; Xcite XYLIS LED fluorescence illumination; Zyla 4.2 Plus sCMOS camera (Andor); and various emission/excitation filter sets.

3 Methods

3.1 Cell maintenance

1. Freshly isolated primary human NK cells are cultured at a density of 1×10^6 cells/mL in RPMI 1640 medium containing 1 ng/mL recombinant human IL-2, 10% heat-inactivated Fetal Calf Serum, 100 U/mL penicillin and 100 μg/mL streptomycin. Only NK cell batches with purity higher than 90% (measured by flow cytometry analysis of CD56 staining) are cultured and used for experiments. NK cells are typically cultured for 3 days before the imaging experiments.
2. Target epithelial cancer cell lines are cultured in appropriate medium supplemented with 10% Fetal Calf Serum, 100 U/mL penicillin and 100 μg/mL streptomycin, and sub-cultured every 2–3 days after cells reach 80–90% confluency using standard Trypsinization method.

3.2 NK cell staining with the LysoBrite red dye

1. Prepare the LysoBrite staining solution by diluting 1 μL of the LysoBrite stock solution into 0.5 mL of Hanks' Buffer with 20 mM Hepes (HHBS) or buffer of your choice (see Note 1).
2. Add 20 μL of the Lysobrite staining solution into 1 mL of NK cells (1×10^6 cells/mL) maintained in the normal NK cell culture medium.
3. Incubate the solution of Lysobrite plus NK cells in the cell culture incubator (37 °C, 5% CO_2) for 1 h.
4. Spin down the LysoBrite-stained NK cells by centrifugation for 8 min at 350 g. Remove the staining solution.
5. Wash the LysoBrite-stained NK cells twice with 8 mL pre-warmed (37 °C) plain RPMI 1640 medium. Each time spin down the NK cells by centrifugation for 8 min at 350 g.
6. Resuspend the washed NK cells at 5×10^5 cells/mL in the live-cell imaging medium (i.e., CO_2-independent medium supplemented with 10% heat-inactivated Fetal Calf Serum, 100 U/mL penicillin, 100 μg/mL streptomycin and proper concentration of IL-2)
7. Incubate the stained NK cells in the cell culture incubator (37 °C, 5% CO_2) for 30 min, which allows the NK cells to recover their healthy polarized morphology. The LysoBrite-stained NK cells are now ready to be used in the live-cell imaging experiments.

3.3 Generation of stable fluorescent reporter cancer cell lines expressing the Granzyme B-FRET reporter and/or IMS-RP (mitochondria reporter)

1. 293 T cells are used to produce retroviral particles that carry the respective fluorescent reporter construct for infecting the target cancer cell lines. On day 1, split one 10 cm dish of confluent 293 T cells at 1:10 ratio into a new 10 cm dish. At the time of vector transfection on day 2, the 293 T cells should reach about 50% confluency.
2. On day 2, change the culture medium of 293 T cells to 9 mL of fresh DMEM medium 2–3 h before the vector transfection.
3. Warm the X-tremeGene HP DNA transfection reagent, plasmids and plain DMEM medium (no serum, no antibiotics) to room temperature. Briefly vortex the X-tremeGene HP DNA transfection reagent vial.
4. Prepare the transfection complex by diluting the plasmids in plain DMEM medium. The amount of each component in the transfection complex is listed below in Table 1.
5. Add the X-tremeGene HP DNA transfection reagent directly into the transfection mixture containing the plasmids (see Note 2). Mix gently by tapping the tube.
6. Incubate the transfection complex for 15 min at room temperature.
7. Add the transfection complex to the 293 T cells in a dropwise manner. Gently swirl the 10-cm cell culture dish.
8. Incubate the transfected 297 T cells (37 °C, 5% CO_2) for 2 days.
9. On day 4, plate the target cancer cells for retroviral infection at 1×10^5 cells/well in 6-well plates.
10. Collect the virus supernatant from the transfected 293 T cells using 10 mL syringe. Filter the virus supernatant through a 0.45 μm sterile syringe filter into a 15 mL or 50 mL conical-bottom centrifuge tube.
11. Add 1 volume of cold PEG-it Virus Precipitation Solution (4 °C) to every 4 volumes of virus supernatant.
12. Refrigerate the mixture of virus supernatant plus PEG-it solution overnight (at least 12 h) at 4 °C.
13. On day 5, centrifuge the mixture of virus supernatant plus PEG-it solution at 1500 g for 30 min at 4 °C. After centrifugation, the virus particles should appear as a white pellet at the bottom of the centrifuge tube.

Table 1 Components of the transfection mix for 10-cm cell culture dish.

Plain DMEM Medium		1000 μL
Plasmids	Granzyme-B FRET or IMS-RP plasmid	5 μg
	Retrovirus packaging vector	5 μg
X-tremeGene HP DNA transfection reagent		30 μL

14. *Re*-suspend the virus pellet in DMEM medium so that the final volume is 1/10 of the original volume before centrifugation.
15. Add polybrene to the concentrated virus solution at a final concentration of 10 µg/mL.
16. Remove the culture medium of the cancer cells seeded in the 6-well plates and add 1 mL of the concentrated virus solution to each well to infect the cancer cells.
17. Incubate the cancer cells in the virus solution (37 °C, 5% CO_2) for 4 h.
18. Remove the virus solution and add fresh culture medium to the cancer cells.
19. On day 6, add the appropriate selection antibiotic (e.g., 1–5 µg/mL puromycin or 200–500 µg/mL G418) to the cell culture medium.
20. Change to fresh antibiotic medium every 2 days until the selection process is completed. During the selection process, split the cancer cells at 1:2 to 1:4 ratio whenever the cells reach about 85% confluency.

3.4 Transient expression of the Caspase 8-FRET fluorescent reporter in cancer cells

1. Plate the cancer cells at $1–1.8 \times 10^5$ cells/well in 12-well plate. At time of the transfection, cancer cells should reach about 90% confluency.
2. On the day of vector transfection, change the culture medium of the cancer cells to 0.9 mL of fresh culture medium.
3. Warm the X-tremeGene HP DNA transfection reagent, plasmids and plain culture medium (no serum, no antibiotics) to room temperature. Briefly vortex the X-tremeGene HP DNA transfection reagent vial.
4. Prepare the transfection complex by diluting the plasmid in plain DMEM medium. The amount of each component in the transfection mix is listed below in Table 2.
5. Add the X-tremeGene HP DNA transfection reagent directly into the transfection mixture containing the plasmid (see Note 2). Mix gently by tapping the tube.
6. Incubate the transfection complex for 15 min at room temperature.
7. Add the transfection complex to the cancer cells in a dropwise manner. Gently swirl the 12-well plate.
8. Incubate the transfected cancer cells (37 °C, 5% CO_2) for 2 days.
9. The cancers cells transiently expressing the Caspase-8 FRET reporter are now ready to be used in the imaging experiments (see Note 3).

Table 2 Components of the transfection mix for transient transfection in 12-well plate.

Plain culture Medium	100 µL
Caspase 8-FRET plasmid	1 µg
X-tremeGene HP DNA transfection reagent	3 µL

3.5 Preparation of NK-cancer cell co-culture for the imaging experiments

1. 1 or 2 days before the live-cell imaging experiment, seed the epithelial cancer cells in the 96-well imaging plate at the cell density listed below in Table 3.
2. Change the culture medium to 120 μL/well CO_2-independent medium supplemented with 10% heat-inactivated Fetal Calf Serum, 100 U/mL penicillin, 100 μg/mL streptomycin and appropriate concentration of human recombinant IL-2.
3. Depending on the NK-to-cancer cell ratio that the imaging experiment aims to study, add 80 μL/well of the appropriate number of NK cells to each well containing the cancer cell lines (see Notes 4 & 5). We typically use the NK-to-cancer cell ratio of 3:1.
4. Gently pipette up and down the mixture a few times to ensure NK cells are homogeneously distributed in the imaging well. The NK-cancer cell co-culture is now ready for the imaging experiments.

3.6 Phase-contrast and fluorescence live-cell imaging

1. NK cells and cancer cells can be easily distinguished and dynamically followed by morphological tracking based on phase-contrast imaging. Below are the typical settings of the Nikon Eclipse Ti2 inverted microscope system that we use to image NK-cancer cell interaction dynamics by phase-contrast imaging (Table 4).
2. The cytotoxic modes of NK cells toward different cancer cell types and the associated rate-limiting kinetics can be measured by imaging the fluorescent reporters, such as the Granzyme-B FRET reporter, Caspase-8 FRET reporter and IMS-RP, as well as the LysoBrite Red dye that specifically marks the lytic granules of NK cells. Below are the typical system setting that we use to image NK-cancer cell interaction dynamics by fluorescence imaging (see Note 6) (Tables 5–7).

Table 3 Target cancer cell density in 96-well plate for performing the NK-cancer cell co-culture imaging experiments.

Epithelial cancer cell lines	Seeding density (cells/well)	Cell density on the day of imaging experiment		
		Confluency (%)	Estimated cells/well	Cell number in the imaging field of view with a 20× lens
MCF7	0.6×10^4	60	1.3×10^4	70–80
SMMC-7721	0.7×10^4	60	1.3×10^4	70–80
U-2 OS	0.6×10^4	60	1×10^4	50–60
HeLa	0.55×10^4	60	1×10^4	50–60

Table 4 Settings for phase-contrast imaging.

Objective lens	S Plan Fluor ELWD 20×	
Bright Field Channel	Camera Binning: DIA Illuminator Intensity: Exposure Time: Imaging frame rate:	1 × 1 50% 100 ms 4 min/frame

Table 5 Imaging settings for the Granzyme-B FRET and Caspase-8 FRET reporter.

Objective lens	Plan Apo 20× (NA = 0.75)	
CFP FRET Channel (Filters: EX436/20, DM455, EM480/40) **YFP FRET Channel** (Filters: EX436/20, DM455, EM530/30)	*Camera binning: 3 × 3* ND Shutter: ND8 X-Cite Illuminator Intensity: 5% Exposure Time: 120–150 ms Frame rate: 4 min/frame	*Camera binning: 1 × 1* ND Shutter: ND8 X-Cite Illuminator Intensity: 70% Exposure Time: 80 ms Frame rate: 30 s/frame

Table 6 Imaging settings for the IMS-RP reporter.

Objective lens	Plan Apo 20× (NA = 0.75)	
TxRed Channel (Filters: EX560/40, DM595, EM630/60)	Camera Binning: ND Shutter: X-Cite Illuminator Intensity: Exposure Time: Frame rate:	3 × 3 ND8 20% 150–200 ms 4 min/frame

Table 7 Imaging settings for the LysoBrite-stained NK cells.

Objective lens	Plan Apo 20× (NA = 0.75)	
TxRed Channel (Filters: EX560/40, DM595, EM630/60)	Camera Binning: ND Shutter: X-Cite Illuminator Intensity: Exposure Time: Frame rate:	1 × 1 ND8 70% 80 ms 30 s/frame

3.7 Quantification of the NK-cancer cell interaction dynamics based on the imaging data

1. Typical phase-contrast images of primary NK cells in co-culture with epithelial cancer cells, such as U-2 OS, are shown in Fig. 1. The small, suspending NK cells with a polarized head-tail morphology are easily distinguishable from the large, adherent epithelial cancer cells.

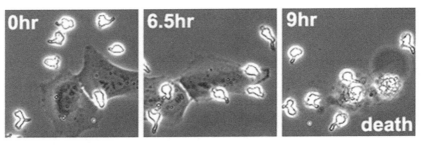

FIG. 1

A representative time sequence of phase-contrast images of human primary NK cells in co-culture with a human cancer cell line, U-2 OS, acquired from live-cell imaging. NK cells were added at time 0 (time is indicated in unit of hours at the upper corner of the still images) and target cancer cell death induced by NK cells can be scored by lysis of the large cancer cells as shown in the last frame of the image sequence.

This data was published in Zhu, Y., Huang, B., Shi, J. (2016). Fas ligand and lytic granule differentially control cytotoxic dynamics of natural killer cell against cancer target. Oncotarget 7, 47163–47172.

2. Based on the image sequence, a number of key dynamic parameters underlying NK-cancer cell interaction can be measured and quantified, including: (1) the total number of target cancer cells killed by the NK cells as a function of time. The number of surviving live cancer cells can then be ratioed to the total number of live cancer cells at time 0 and plotted as cumulative survival curves as a function of time (Fig. 2A). Intuitively, the faster the survival curves decay in time, the more efficiently NK cells kill the cancer targets. This provides a direct measure of the sensitivity of different cancer cell types to NK cell cytotoxic activity, and can be compared with data obtained from ensemble cell viability assays (Zhu et al., 2021). (2) the average number of NK-cancer cell contacts (scored by spatial co-localization of the NK cells and cancer cell based on the imaging data) per cancer cell per unit time (Fig. 2B). This provides a measure of how frequently NK cells directly interact with a cancer target (Zhu et al., 2016). (3) the average duration of the NK-cancer cell contact (i.e., the time for which the co-localization of NK cell and target cell persists) (Fig. 2C), revealing whether sustained or transient NK-cancer cell interaction is involved. (4) Motility of NK cells toward the cancer targets. Based on the phase-contrast morphology of NK cells, they can be tracked by established single-cell tracking algorithms, and the migration behaviors of NK cells, e.g., chemotaxis, can be quantified.

3. We typically perform three independent imaging experiments and average the single cell statistics from the three experiments to obtain the mean. For each treatment condition in the imaging experiment, we find analysis of 50 to 100 single cells (either NK cells or target cells, depending on the phenotypes to be investigated) is sufficient to obtain single cell statistics that are representative of the population average as well as the variability. Error bars, e.g., shown in Fig. 2B and C, are standard deviations of the three independent imaging experiments. Statistical analysis, e.g., P value, is usually obtained by student's t-test.

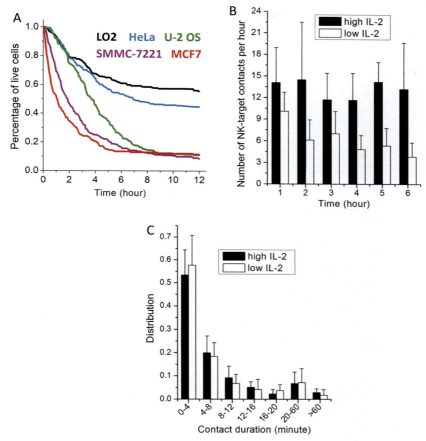

FIG. 2

(A) Cumulative survival curves of different epithelial cell lines, including LO2 (denoted in black), HeLa (blue), U-2 OS (green), SMMC-7721 (magenta) and MCF7 (red), in co-culture with primary human NK cells. (B) Average number of contacts per hour and (C) average contact duration between a target U-2 OS cell and NK cells under high IL-2 (50 ng/mL) vs. low IL-2 (0.2 ng/mL).

These data were published in Zhu, Y., Huang, B., Shi, J. (2016). Fas ligand and lytic granule differentially control cytotoxic dynamics of natural killer cell against cancer target. Oncotarget 7, 47163–47172; Zhu, Y., Xie, J., Shi, J. (2021). Rac1/ROCK-driven membrane dynamics promote natural killer cell cytotoxicity via granzyme-induced necroptosis. BMC Biology 19(1), 140.

4. To study the rate-limiting kinetics underlying different NK cell cytotoxic modes, e.g., lytic granule-mediated apoptosis, death ligand-mediated apoptosis and necrosis, the Granzyme B-FRET reporter or the Caspase 8-FRET reporter in combination with the mitochondria reporter, IMS-RP, can be employed.

60 CHAPTER 4 Cytotoxic and chemotactic dynamics of NK cells

For instance, Fig. 3 shows the signature fluorescence signals from the Granzyme B-FRET reporter and IMS-RP upon target cancer cell death triggered by primary NK cells (Zhu et al., 2021). The upper image panels in color are fluorescence images of the Granzyme B-FRET reporter and the lower image panels in gray scale are the IMS-RP signals from SMMC-7721, U-2 OS and MCF7, respectively. The Granzyme B-FRET images are overlay of the CFP (denoted by blue) and YFP (green) channels. CFP and YFP signals from a single target cell can be quantified from the image sequence, e.g., using ImageJ, by first manually

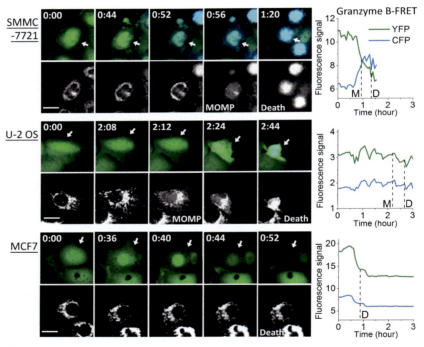

FIG. 3

Left panels: Representative fluorescence image sequence of the Granzyme B-FRET reporter (overlay of the CFY (blue) and YFP (green) signal) and the IMS-RP reporter (gray signal) expressed in the target cancer cells upon interaction with primary NK cells. Time (unit: hour: minute) is indicated at the top left corner of each Granzyme B-FRET image. The white scale bar is 20 μm and the white arrows point to the specific target cancer cells, for which the CFP and YFP fluorescence signals were quantified and shown in the right panels. Right panels: Single cell trajectories of CFP and YFP fluorescence signals quantified from the time-lapse movies using ImageJ. The time of MOMP (scored by the IMS-RP signal abruptly changing from punctate to diffused localization) and the time of death (scored morphologically by cell blebbing and lysis) are indicated by the vertical dotted line.

These data were published in Zhu, Y., Xie, J., Shi, J. (2021). Rac1/ROCK-driven membrane dynamics promote natural killer cell cytotoxicity via granzyme-induced necroptosis. BMC Biology 19(1), 140.

segmenting the single target cell and then summing up the fluorescence signal from the CYP and YFP channel, respectively. Representative CFP and YFP signals from a single target cell are shown next to the image panels in Fig. 3. As discussed above in the Materials section, if target cancer cells die due to granzyme B activity from NK cells, such death is preceded by a loss of the Granzyme B-FRET (i.e., an increase of CFP signal and decrease of YFP signal) followed by mitochondrial outer membrane permeabilization (MOMP) that can be scored by the IMS-RP signal abruptly changing from punctate to diffused localization (refer to SMMC-7721 in Fig. 3). For cell death without a change in the Granzyme B-FRET signal but with MOMP, it is most likely induced by death ligand-mediated extrinsic apoptosis (refer to U-2 OS in Fig. 3). The necrotic killing mode of NK cells can also be detected by an abrupt loss of the Granzyme B-FRET reporter in the target cancer cells due to membrane ruptures as well as a lack of MOMP (refer to MCF7 in Fig. 3).

5. Once the specific NK cell killing modes are determined by the above fluorescence signatures, kinetics associated with the different killing modes can be individually quantified using similar rationales discussed above in point #2 for phase-contrast imaging data.

6. A critical dynamic step involved in lytic granule-mediated NK cell killing is translocation of the lytic granules originally stored in the tails of NK cells to the cytoplasm of the target cancer cells. The rate-limiting phenotypes and kinetics involved in the lytic granule translocation can be conveniently monitored and measured by imaging the LysoBrite dye that marks the acidic lytic granules of NK cells. As shown in Fig. 4A, lytic granules are stored in NK cells' tails, as NK

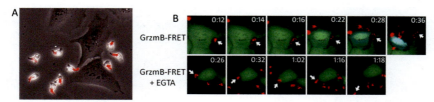

FIG. 4

(A) Lytic granules of primary NK cells can be probed and imaged by the acidic granule marker, LysoBrite. The image shown is the overlay of the phase-contrast image and the LysoBrite image (denoted in red). (B) Representative sequences of the dynamic movement of lytic granules upon formation of the NK-cancer cell immunological synapse. The images were overlay of the CFP (blue) and YFP (green) fluorescence from the Granzyme B-FRET reporter expressed in U-2 OS cells and the red fluorescence from LysoBrite in NK cells. Time is indicated in the unit of hour: minute at the top right corner of each image. The specific NK cells that interact and form conjugation with the target U-2 OS cells are indicated by the white arrows. EGTA was used at 0.8 mM.

These data were published in Zhu, Y., Huang, B., Shi, J. (2016). Fas ligand and lytic granule differentially control cytotoxic dynamics of natural killer cell against cancer target. Oncotarget 7, 47163–47172.

cells move and transiently interact with the target cancer cells. Upon recognition of a cancer target and formation of the immunological synapse (IS), the lytic granules move from the tail to the front of NK cell, and then disperse at the IS followed by translocation to the cancer cell cytoplasm, which led to the activation of granzyme B activity (scored by the loss of Granzyme B-FRET) (Fig. 4B) (Zhu et al., 2016). EGTA is a calcium chelator and known to inhibit lytic granule transfer. Treatment of EGTA indeed prevents lytic granule dispersion at the IS and the subsequent transfer into the target cancer cells, despite that NK cells are still able to recognize and form sustained conjugation with the target cancer cells (Fig. 4B). Based on our experiences, the LysoBrite red dye has minimal toxicity to primary NK cells and is much more photostable than other live-cell dyes that mark the acidic granules, such as the LysoTracker.

4 Notes

1. The stock solution of LysoBrite red dye obtained from AAT Bioquest does not specify the concentration of the dye. We store the stock solution at $\leq -20\,°C$ in small aliquots (8 µL/vial) to avoid repeated freeze-thaw cycles. Both the LysoBrite stock solution and the medium for stock dilution should be warmed up to room temperature before the staining procedures.
2. The X-tremeGENE HP DNA Transfection Reagent should be pipetted directly into the medium containing the plasmids without coming into contact with the walls of the plastic tube.
3. We used stable transfection of the Granzyme B-FRET reporter and IMS-RP, while transient transfection of the Caspase 8-FRET reporter, purely out of convenience. As the Caspase 8-FRET reporter that we have in hand is in a regular vector, not a viral vector, it will take much longer time to generate stably transfected cells using this plasmid. Therefore, we simply used transient transfection of the Caspase 8-FRET reporter to measure the activation dynamics of caspase 8 induced by NK cells.
4. To prepare the NK cell stock solution to add to the cancer cell culture, collect the primary NK cells from the cell culture dish or wells and centrifuge them down at 350 g for 8 min. Remove the excessive medium and then re-suspend the NK cells at the desired concentration in the live-cell imaging medium.
5. After the centrifugation step(s), we find the NK cells tend to lose their polarized morphology and appear to be stressed. By putting the NK cells back to the incubator (37 °C, 5% CO_2) and let them recuperate for 30 min, the NK cells generally recover their healthy polarized morphology. We thus recommend to put the NK cells in the incubator for 30 min after procedures involving centrifugation and before performing the imaging experiments.
6. The fluorescence reporters can be imaged at high spatiotemporal resolution (i.e., with no camera pixel binning (1 × 1) and a fast imaging frame rate of 30 s/frame),

or medium spatiotemporal resolution (i.e., with camera pixel binning (3 × 3) and an imaging frame rate of 4 min/frame). As expected, the fluorescence reporters photobleach much faster under the imaging settings of high spatiotemporal resolution. Hence, the total viable time for imaging at high spatiotemporal resolution is typically less than 3 h, while more long-term live-cell imaging, i.e., 12–24 h, can be performed for medium spatiotemporal resolution. For measuring the cytotoxic activities of NK cells, we typically use the medium resolution settings and image for long duration, while for imaging the lytic granule dynamics using LysoBrite, we generally use the high-resolution settings and image for about 2–3 h.

Acknowledgments

We thank Dr. Paul Choi and Dr. Timothy Mitchison (Department of Systems Biology, Harvard Medical School) for the Granzyme B-FRET retroviral reporter construct, and Dr. John Albeck and Dr. Peter Sorger (Department of Systems Biology, Harvard Medical School) for the Caspase 8-FRET reporter and the IMS-RP retroviral vector. The work described and discussed in this article was supported by funding from the Hong Kong Research Grant Council (#C2006-17E, #T12-710/16-R and #12302421) to J. Shi and the "Laboratory for Synthetic Chemistry and Chemical Biology" under the Health@InnoHK Program by the Innovation and Technology Commission of Hong Kong.

References

Albeck, J. G., Burke, J. M., Aldridge, B. B., Zhang, M., Lauffenburger, D. A., & Sorger, P. K. (2008). Quantitative analysis of pathways controlling extrinsic apoptosis in single cells. *Molecular Cell, 30*(1), 11–25.

Bhat, R., & Watzl, C. (2007). Serial killing of tumor cells by human natural killer cells--enhancement by therapeutic antibodies. *PLoS One, 2*(3), e326.

Bryceson, Y. T., March, M. E., Ljunggren, H. G., & Long, E. O. (2006). Activation, coactivation, and costimulation of resting human natural killer cells. *Immunological Reviews, 214*, 73–91.

Caligiuri, M. A. (2008). Human natural killer cells. *Blood, 112*(3), 461–469.

Champsaur, M., & Lanier, L. L. (2010). Effect of NKG2D ligand expression on host immune responses. *Immunological Reviews, 235*(1), 267–285.

Chiossone, L., Dumas, P. Y., Vienne, M., & Vivier, E. (2018). Natural killer cells and other innate lymphoid cells in cancer. *Nature Reviews. Immunology, 18*(11), 671–688.

Choi, P. J., & Mitchison, T. J. (2013). Imaging burst kinetics and spatial coordination during serial killing by single natural killer cells. *Proceedings of the National Academy of Sciences of the United States of America, 110*, 6488–6493.

Drake, C. G., Jaffee, E., & Pardoll, D. M. (2006). Mechanisms of immune evasion by tumors. *Advances in Immunology, 90*, 51–81.

Lanier, L. L. (2005). NK cell recognition. *Annual Review of Immunology, 23*, 225–274.

Lee, B. J., & Mace, E. M. (2017). Acquisition of cell migration defines NK cell differentiation from hematopoietic stem cell precursors. *Molecular Biology of the Cell*, *28*(25), 3573–3581.

Lee, S. H., Miyagi, T., & Biron, C. A. (2007). Keeping NK cells in highly regulated antiviral warfare. *Trends in Immunology*, *28*(6), 252–259.

Olofsson, P. E., Forslund, E., Vanherberghen, B., Chechet, K., Mickelin, O., Ahlin, A. R., et al. (2014). Distinct migration and contact dynamics of resting and IL-2-activated human natural killer cells. *Frontiers in Immunology*, *5*, 80.

Prager, I., & Watzl, C. (2019). Mechanisms of natural killer cell-mediated cellular cytotoxicity. *Journal of Leukocyte Biology*, *105*(6), 1319–1329.

Vivier, E., Tomasello, E., Baratin, M., Walzer, T., & Ugolini, S. (2008). Functions of natural killer cells. *Nature Immunology*, *9*(5), 503–510.

Zhou, Z., He, H., Wang, K., Shi, X., Wang, Y., Su, Y., et al. (2020). Granzyme A from cytotoxic lymphocytes cleaves GSDMB to trigger pyroptosis in target cells. *Science*, *368*(6494), eaaz7548.

Zhu, Y., Huang, B., & Shi, J. (2016). Fas ligand and lytic granule differentially control cytotoxic dynamics of natural killer cell against cancer target. *Oncotarget*, *7*, 47163–47172.

Zhu, Y., Xie, J., & Shi, J. (2021). Rac1/ROCK-driven membrane dynamics promote natural killer cell cytotoxicity via granzyme-induced necroptosis. *BMC Biology*, *19*(1), 140.

CHAPTER

Quantification of interaction frequency between antigen-presenting cells and T cells by conjugation assay

Ondrej Cerny*
Institute of Microbiology of the Czech Academy of Sciences, Prague, Czech Republic
*Corresponding author: e-mail address: ondrej.cerny@biomed.cas.cz

Chapter outline

1 Introduction..66
2 Materials...67
3 Methods..68
4 Concluding remarks...72
5 Notes..72
Acknowledgment..74
References...74

Abstract

Interaction between an antigen-presenting cell and a T cell, and their subsequent conjugation are a prerequisite for the formation of the immunological synapse and productive, antigen-dependent activation of T cells. This initial interaction is accompanied by recognition of the presented antigen by the T cell receptor, and by changes in the morphology of the interacting cells and in actin cytoskeleton structure in the site of interaction. The experimental protocol below describes a simple assay for quantitative assessment of antigen-presenting cells-T cell conjugation using confocal microscopy or flow cytometry.

1 Introduction

Effective activation of T cells in an antigen-dependent manner requires establishment of the immunological synapse (IS), a contact site between an antigen-presenting cell (APC) and the responding T cell (nicely reviewed by Dustin, 2014). Following the initial contact between the APC and the T cell, three major signaling events between the cells takes place at the IS. The peptide-loaded major histocompatibility complex (MHC) interacts with the T cell receptor (TCR) complex in the IS. Co-stimulatory signals represented by B7 molecules (CD80 and CD86) on the APC needed for T cell activation are also provided in the IS (Viola, Schroeder, Sakakibara, & Lanzavecchia, 1999). Furthermore, the IS is stabilized by interactions between adhesion molecules from both cells involved in the IS. This allows sufficient time for integration of transferred signals by both involved cells (Martin-Cofreces, Vicente-Manzanares, & Sanchez-Madrid, 2018).

After the initial interaction between an APC and a T cell, the T cell halts at the APC. This is typically followed by distinct morphological changes leading to a dramatic increase in the contact area between the two interacting cells. Then, the MHC complex is redistributed to the APC-T cell interaction interphase (Wetzel, McKeithan, & Parker, 2002). This is accompanied by a dramatic increase in phosphotyrosine levels in the T cell side of the interaction interphase (Wetzel et al., 2002) caused by the signaling of TCR and precedes the complete formation of the IS (Lee et al., 2002). The initial interaction of the APC and the T cell before the formation of the mature IS is termed "conjugation."

Formation of the IS takes only minutes and is accompanied by a rearrangement of the cell cytoskeleton, including accumulation of filamentous actin under the plasma membrane in the IS (Grakoui et al., 1999; Wetzel et al., 2002). While the cell-cell contact may last for prolonged time periods (Miller, Wei, Parker, & Cahalan, 2002; Stoll, Delon, Brotz, & Germain, 2002) or be short-lived and sequential (Gunzer et al., 2000), the architecture of IS remains flexible throughout the whole interaction (Sims et al., 2007).

The mature IS is organized into distinct structures termed supramolecular activation complexes (SMACs) (Monks, Freiberg, Kupfer, Sciaky, & Kupfer, 1998). Distinct protein functions are spatially segregated into central SMAC (cSMAC), peripheral SMAC (pSMAC) and distal SMAC (dSMAC). The organization of the SMAC structure depends largely on the actin cytoskeleton (Bunnell, Kapoor, Trible, Zhang, & Samelson, 2001; Campi, Varma, & Dustin, 2005; Comrie, Babich, & Burkhardt, 2015; Comrie, Li, Boyle, & Burkhardt, 2015; Stinchcombe, Majorovits, Bossi, Fuller, & Griffiths, 2006). Therefore, the accumulation of actin cytoskeleton may be used as a marker of the IS formation.

Here, a detailed protocol for quantification of antigen-specific interaction events between APCs and T cells using either confocal microscopy or flow cytometry is presented. The described assay is based on a similar flow cytometry assay (Comrie, Li, et al., 2015), but simplified and adapted for confocal microscopy. The protocol below offers a simple and easy-to-do assay feasible in most laboratory settings.

2 Materials

(a) Cell culture handling and labeling
  (i) Cell lines
  - MutuDC cell line (Fuertes Marraco et al., 2012)
  - B3Z T cell hybridoma (Karttunen, Sanderson, & Shastri, 1992)

  (ii) Cell line media
  - Iscove's modified Dulbecco's medium (IMDM-glutamax; GIBCO #31980)
  - RPMI 1640 (Invitrogen #31870–074)
  - Fetal calf serum, heat-inactivated (GIBCO; see Note 1)
  - β-Mercaptoethanol (GIBCO #31350)
  - HEPES pH 7.4, TC grade (GIBCO #15630)
  - 1 × PBS pH 7.4

  (iii) Cell treatment and fixation
  - *Salmonella typhimurium* LPS (Sigma-Aldrich #L6143-1mg)
  - Model antigen (e.g., SIINFEKL (chicken Ovalbumin peptide (aa 257–264); Sigma-Aldrich # S7951-1MG)
  - 6% PFA in 1 × PBS pH 7.4

  (iv) Labeling
  - Triton X-100 (Sigma #T9284)
  - CellTracker Blue CMF2HC Dye (Thermo Fisher Scientific #C12881)
  - Alexa Fluor 555 Phalloidin (Thermo Fisher Scientific #A34055)
  - Mounting solution for confocal microscopy (e.g., ibidi Mounting Medium #50001)

  (v) Plastic, material and instruments
  - Tissue culture dishes
  - 50 mL centrifugation tubes
  - 96-well TC-treated plates, U bottom
  - 6-well TC-treated plates
  - Glass coverslips for confocal microscopy (e.g., Marienfeld, no. 1.5 cat. no. 0117530)
  - 24-well TC-treated plates
  - 10 mL serological pipettes and automatic pipette
  - Micropipettes
  - Multichannel pipette
  - Hemocytometer
  - Humidified incubator at 37 °C, 5% CO_2 (PHCbi)
  - Inverted microscope (Olympus, CKX53)
  - Benchtop centrifuge for 50 mL tubes (Thermo Fisher Scientific)
  - Benchtop centrifuge for 96-well plates (Thermo Fisher Scientific)
  - Vortex for 96-well plates (Thermomixer Comfort, Eppendorf)

(b) Confocal imaging
  - Inverted confocal microscope (e.g., LSM 710, Olympus)

(c) Flow cytometry
- Flow cytometer (e.g., LSR Fortessa, BD)

(d) Data analysis
- Image analysis software (e.g., freeware ImageJ (https://imagej.net) or CellProfiler (https://cellprofiler.org/))
- Flow cytometry data analysis software (e.g., FlowJo, BD)

3 Methods

For successful completion and analysis of the assay, several controls are crucial and preparation of extra cells for these controls is necessary. More specifically, samples containing one cell type only (a single-labeled control for fluorescence signal detection and a control of scatter set up for flow cytometry), a sample without antigen peptide (a control for determination of unspecific cell crosslinking during fixation) and a sample with nonactivated APCs shall be included.

(a) Cell culture and handling

1. Mouse dendritic cell (DC) line MutuDC1940 (see Note 2) express low levels of cytosolic GFP. This cell line is routinely cultured in IMDM (IMDM-glutamax; GIBCO #31980) supplemented with 10% FCS, 10 mM HEPES pH 7.4 (GIBCO #15630) and 50 µM β-mercaptoethanol (GIBCO #31350) and kept in humidified incubator at 37 °C and 5% CO2. Split the cells before they start growing in colonies and clumps, avoid excessive mechanical stress.
    i. To split or harvest MutuDCs, incubate the cells in non-enzymatic cell dissociation buffer (5 mM EDTA, 20 mM HEPES in PBS pH 7.4) for 5–10 min at 37 °C and 5% CO_2.
    ii. Wash the plate with cell dissociation buffer and centrifuge the detached cells at $300 \times g$ for 5 min.
    iii. Resuspend the cells in fresh IMDM and seed the cells into an appropriate flask. The doubling time of MutuDCs is around 24 h and standard subculture should be in ratios between 1:4 and 1:10. The cells will, however, grow even at very high dilutions, although the growth time may prolong. Optimal cell density is between 0.2 and 3×10^5 cells/cm^2.
2. B3Z T cell hybridoma (see Note 3) is routinely cultured in RPMI 1640 (Invitrogen #31870–074) supplemented with 10% FCS, 10 mM HEPES pH 7.4 (GIBCO #15630), 1 mM Na pyruvate, 2 mM Glutamine and 50 µM β-mercaptoethanol (GIBCO #31350).
    i. B3Zs are non-adherent cells and are split by moving a fraction of cells to a new plate.
    ii. To split the B3Z T cells, harvest the medium containing suspension cells without removing the adherent fraction of cells from plastic and centrifuge the detached cells at $300 \times g$ for 5 min.

iii. Resuspend the cells in fresh RPMI 1640, count in hemocytometer and estimate the viability using trypan blue exclusion. Move a fraction into a new plate. The doubling time of B3Z T cells is less than 24 h and standard subculture should be in ratios between 1:5 and 1:20. Optimal cell density is between 0.2 and 3×10^5 cells/cm^2.

(a) MutuDC preparation
  A. For confocal microscopy
   1. One day before the experiment, place cleaned and autoclaved (sterile) glass coverslips (13 mm) into wells in a TC-treated 24-well plate.
   2. Harvest and count MutuDCs.
   3. Seed 1×10^5 MutuDCs in 1 mL of IMDM supplemented with 100 ng/mL LPS (in PBS) (see Note 4)
   4. Incubate o/n (see Note 5) at 37 °C and 5% CO_2.
   5. Replace the spent medium with 300 µL of fresh IMDM with 1 µg/mL of SIINFEKL peptide; incubate for 1 h (see Note 6). Keep a sample without SIINFEKL as a negative control (see Note 7).
   6. Wash with 1 mL of PBS and keep in 1 mL of complete RPMI until labeled B3Z can be added. Remove the medium before adding labeled B3Z T cells (see Note 8).
  B. For flow cytometry
   1. One day before the experiment, activate 1×10^6 MutuDCs in 2 mL of IMDM supplemented with 100 ng/mL of LPS (in PBS) (see Note 4) in a 6-well plate.
   2. Incubate o/n at 37 °C and 5% CO_2.
   3. On the day of the experiment, harvest and count MutuDCs.
   4. Seed 3×10^5 MutuDCs in 100 µL of IMDM supplemented with 1 µg/mL SIINFEKL peptide into one well in a U-bottom 96-well plate, incubate for 1 h (this time is not sufficient for adhesion of MutuDCs to plastic surface). Keep a sample without SIINFEKL as a negative control.
   5. Wash with 100 µL of PBS (centrifugation at 300 g for 3 min is needed to allow removal of the supernatant without losing non-adherent MutuDCs) and keep in 100 µL of complete RPMI until labeled B3Z can be added (see Note 8). Remove the medium before adding labeled B3Z T cells. This has to be done before MutuDC adhere to the plastic surface. Mix well.

(b) B3Z T cell labeling
   1. On the day of the experiment, harvest, wash with PBS, and count B3Z T cells.
   2. Prepare B3Z T cells in 50 mL centrifugation tubes at a concentration 1 million cells per ml in PBS (see Note 9).
   3. To label B3Z T cells, add 1 µL of CellTracker Blue per 1 million cells (see Note 10). Vortex immediately.
   4. Incubate for 10 min at 37 °C in the dark.

5. Wash labeled B3Z T cells three times with 20 mL of PBS (see Note 11).
6. Count the labeled B3Z T cells during the last washing step.
7. Resuspend labeled B3Z T cells in complete RPMI at the concentration needed for the conjugation assay (see Note 12). Use 4×10^5 or 6×10^5 B3Z T cells (B3Z to MutuDCs ratio 2:1) in (A) 300 μL of RPMI for confocal microscopy or (B) 100 μL of RPMI for flow cytometry, respectively.

(c) Conjugation assay
1. Add labeled and washed B3Z T cells to SIINFEKL-pulsed and washed MutuDCs.
2. (A) Centrifuge the 24-well plate for 5 min at $90 \times g$ to sediment the B3Z T cells.
   (B) Mix gently the B3Z T cells with MutuDCs using a multichannel pipette.
3. Incubate for 1 h at 37 °C and 5% CO_2 in the dark (see Note 13).
4. Vortex at 1000 RPM for 5 s to separate the loosely attached cells.
5. Add pre-warmed paraformaldehyde into the final concentration of 3% (see Note 14) to cross-link the conjugates, mix gently and incubate for 10 min at 37 °C in the dark.
6. Wash three times with PBS (see Note 15).
7. (A) For confocal microscopy, permeabilise fixed cells with 0.1% Triton X-100 for 5 min at RT in the dark.
   (B) For flow cytometry, the cells are now ready for analysis.
8. (A) Wash three times with PBS.
9. (A) Label the coverslips with 1 unit of Alexa Fluor 555 Phalloidin for 30 min at RT in the dark.
10. (A) Wash three times with PBS, twice with dH_2O and mount on glass slides for confocal microscopy (see Note 16). **The samples are now ready for analysis**.

(d) Data acquisition and analysis
   A. For confocal microscopy
   1. Set confocal microscope parameters according to the single-labeled controls into the dynamic range. CellTracker Blue signal may be acquired with DAPI settings.
   2. In a negative control (sample without SIINFEKL) and with experimental samples, choose several random fields with low magnification according to the GFP fluorescence (MutuDCs) without monitoring the B3Z T cell or actin channels.
   3. Sample the same positions for B3Z T cell and actin channels as for MutuDC channel.
   4. In an image analysis software (e.g., freeware ImageJ or CellProfiler), count the number of MutuDC-T cell contacts (visualized by actin enrichment on cell-cell contact) and plot as a fraction of all sampled MutuDCs (see Notes 17 and 18, Fig. 1):

3 Methods   71

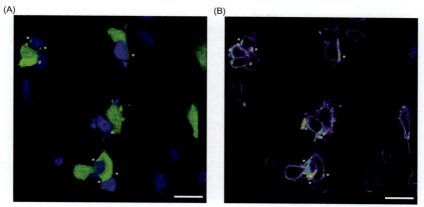

FIG. 1

Conjugation assay performed by confocal microscopy. (A) A representative confocal microscopy image of MutuDCs (green) interacting with B3Z T cells (blue) in SIINFEKL-dependent manner. (B) Actin localization into the areas of MutuDC-B3Z T cell localization from (A). The false colouring is an LUT rainbow2 representation of the signal to noise ratio. *—the areas with increased actin accumulation at the MutuDC-B3Z T cell interaction interphase; #—no actin accumulation at the MutuDC-B3Z T cell interaction interphase; Scale bar—20 μm.

frequency of conjugation

$$= \frac{\text{number of detected actin enrichments between MutuDC and B3Z T cell}}{\text{number of all MutuDCs in all sampled fields of view}}$$

B. For flow cytometry
1. Set flow cytometer parameters according to the unlabeled and single-labeled controls into the dynamic range. CellTracker Blue signal may be acquired with DAPI settings.
2. In a flow cytometry analysis software (e.g., FlowJo), gate on all cells removing debris (see Notes 13). Gate on all GFP$^+$ events in the cell gate. Acquire the same number of events in all samples.
3. Count percentage of GFP$^+$ CellTracker Blue$^+$ events out of all GFP$^+$ events (see Notes 18, 19 and 20, Fig. 2).

$$\textit{frequency of conjugate formation} = \frac{\textit{number of GFP}^+\textit{CellTracker Blue}^+ \textit{ events}}{\textit{number of all GFP}^+ \textit{ events}} \times 100$$

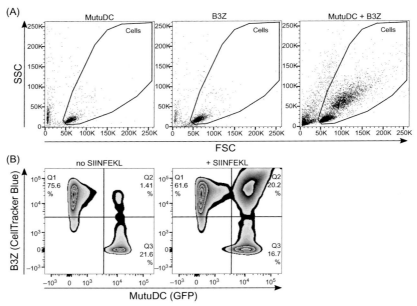

FIG. 2

Example of results of the conjugation assay for flow cytometry. (A) Flow cytometry dot plot showing the distribution of MutuDCs (left), B3Z T cells (middle) and their conjugates (right) on the FCS, SSC plot. The gate "Cells" was used to quantify the composition of conjugates. (B) An example showing the comparison of background antigen-nonspecific PFA-dependent cell crosslinking (left) and antigen-specific conjugate formation.

4 Concluding remarks

The assay described above provides a simple but reliable and quantifiable assessment of the functionality of initial steps of interaction and immunological synapse formation between antigen-presenting cells and T cells. This assay can be done in a basic laboratory setting with no need for advanced instrumentation. However, it can be easily upgraded for more advanced technological solutions such as imaging flow cytometry (Wabnitz, Kirchgessner, & Samstag, 2019). For more recommendations, see the notes below. Use of mutant cells lacking proteins of interest or additional labeling of these proteins on either cell type may be used to examine function of these proteins in the initial phases of interaction between antigen-presenting cells and T cells (Cerny et al., 2021).

5 Notes

1. FCS for MutuDC cultivation shall be tested as some serum batches do not support MutuDC growth (Hans Acha-Orbea, personal communication). Split

the cells before they reach 2×10^6 cells/mL or 20×10^6 cells on a 10-cm TC-treated Petri dish.

2. Other antigen-presenting cells (e.g., bone marrow-derived DCs) may be used. If the chosen cells do not express a fluorescent protein endogenously, they might be labeled before pulsing with the SIINFEKL peptide (steps **3bA5** for confocal microscopy or **3bB4** for flow cytometry) according to the step **3c** with a compatible cell marker (e.g., CellTracker™ Green CMFDA Dye (Thermo Fisher Scientific #C7025) or CFSE (Thermo Fisher Scientific #65-0850-84).
3. Other T cells may be used instead of B3Z T cell hybridoma. CD8+ and CD4+ T cells from OT-I and OT-II mice, respectively, isolated from spleen and mesenteric lymph nodes, provide comparable results to B3Z T cells (Cerny et al., 2021).
4. DC activation is needed to get the maximal capacity to interact with and activate the T cells. Other DC activating signals may be used. Differences between activated and non-activated DCs may be used to monitor variation between experiments.
5. Aim for equally distributed cells not touching each other with enough space between neighboring cells to accommodate the T cells on the day of the experiment.
6. Fluorescent ovalbumin (e.g., Alexa Fluor 647 Conjugate, Invitrogen #O34784) or ANTI MHC-OVA peptide antibody (e.g., OVA257–264 (SIINFEKL) peptide bound to H-2Kb Monoclonal Antibody, Thermo Fisher Scientific #25–5743-82) may be used to check for the efficiency of antigen presentation. SIINFEKL concentration and time needed for sufficient antigen presentation shall be optimized for conditions in the laboratory.
7. Non-specific localization of T cells in the DC vicinity may be identified by monitoring actin accumulation at the immunological synapse
8. Time stability of antigen presentation shall be examined to identify the optimal time window for the addition of T cells.
9. Avoid using FCS as it may interfere with the labeling.
10. Optimal concentration of the label should be determined according to the sensitivity of the fluorescence detectors used during sample acquisition.
11. Alternatively, the complete RPMI medium may again be used now as FCS would quench the excess of the dye.
12. B3Z T cells are prepared in the same way for both confocal microscopy and flow cytometry, and both assays may be done simultaneously. In this case, it is preferential to label more B3Z T cells at once and split them after washing in step **3b7**. This allows for a higher consistency between the experiments.
13. It may be important to include fixable live/dead dye in the step **3c3** (e.g., eFluo 780; Thermo Fisher Scientific #65-0865). The exact dye and its concentration have to be adjusted based on other fluorophores used in the rest of the assay.
14. Avoid unnecessary washing steps to limit temperature changes possibly influencing the actin cytoskeleton.

15. If labeling of CD3 or other surface markers is required, the corresponding primary antibodies may be added after fixation of conjugates in the step **3c6**. If the used antibody is sensitive to paraformaldehyde-induced conformational changes in the fixed sample, it may be added to DC - T cell mixture in step **3c3** 45 min after the mixing of T cells and DCs and incubated with cells for 15 min before vortexing and paraformaldehyde fixation. Binding of the antibodies to living cells may, however, also trigger signaling of the labeled receptor, which may cause some bias in signaling events. Check also for unspecific uptake of these antibodies by DCs.
16. Glass slides supporting adhesion of cells fixed in suspension (e.g., SuperFrost Plus Gold, VWR cat. n. 630-1324) may be used to perform microscopy analyses on conjugates formed in suspension according to the protocol (B).
17. Importantly, one cell (either DC or T cell) can form IS with more cells.
18. This formula gives a percentage of DCs involved in conjugates. If this information is required for the T cells, the denominator could be changed into "number of all B3Z T cells in all sampled fields of view" for microscopic analysis or "number of all CellTracker Blue$^+$ events" for cytometric analysis.
19. In addition to antigen-nonspecific interactions between DCs and T cells, nonspecific PFA-dependent crosslinking between two or more DCs or T cells may be visualized using GFP-A/GFP-H or CellTracker Blue-A/CellTracker Blue-H dot plots in analogy to "singlet" gating.
20. The median fluorescence intensity signal of conjugates in each channel shall be the sum of median fluorescence intensity signals of DCs and T cells. This shall lead to a slight but observable shift in the median fluorescence intensity of conjugates in comparison to single-cell gates.

Acknowledgment

MutuDCs were a gift from Hans Acha-Orbea (University of Lausanne). Special thanks belong to Prof. David Holden in whose laboratory at Imperial College London the protocols were optimized. I would like to thank Dr. Camilla Godlee from Imperial College London for help with confocal microscopy and Dr. Jessica E. Rowley for help with flow cytometry. O.C. is supported by the Czech Science Foundation (22-05356S). O.C. has no competing interests.

References

Bunnell, S. C., Kapoor, V., Trible, R. P., Zhang, W., & Samelson, L. E. (2001). Dynamic actin polymerization drives T cell receptor-induced spreading: A role for the signal transduction adaptor LAT. *Immunity*, *14*, 315–329.

Campi, G., Varma, R., & Dustin, M. L. (2005). Actin and agonist MHC-peptide complex-dependent T cell receptor microclusters as scaffolds for signaling. *The Journal of Experimental Medicine*, *202*, 1031–1036.

Cerny, O., Godlee, C., Tocci, R., Cross, N. E., Shi, H., Williamson, J. C., et al. (2021). CD97 stabilises the immunological synapse between dendritic cells and T cells and is targeted for degradation by the Salmonella effector SteD. *PLoS Pathogens*, *17*, e1009771.

References

Comrie, W. A., Babich, A., & Burkhardt, J. K. (2015). F-actin flow drives affinity maturation and spatial organization of LFA-1 at the immunological synapse. *The Journal of Cell Biology, 208*, 475–491.

Comrie, W. A., Li, S., Boyle, S., & Burkhardt, J. K. (2015). The dendritic cell cytoskeleton promotes T cell adhesion and activation by constraining ICAM-1 mobility. *The Journal of Cell Biology, 208*, 457–473.

Dustin, M. L. (2014). The immunological synapse. *Cancer Immunologic Research, 2*, 1023–1033.

Fuertes Marraco, S. A., Grosjean, F., Duval, A., Rosa, M., Lavanchy, C., Ashok, D., et al. (2012). Novel murine dendritic cell lines: A powerful auxiliary tool for dendritic cell research. *Frontiers in Immunology, 3*, 331.

Grakoui, A., Bromley, S. K., Sumen, C., Davis, M. M., Shaw, A. S., Allen, P. M., et al. (1999). The immunological synapse: A molecular machine controlling T cell activation. *Science, 285*, 221–227.

Gunzer, M., Schafer, A., Borgmann, S., Grabbe, S., Zanker, K. S., Brocker, E. B., et al. (2000). Antigen presentation in extracellular matrix: Interactions of T cells with dendritic cells are dynamic, short lived, and sequential. *Immunity, 13*, 323–332.

Karttunen, J., Sanderson, S., & Shastri, N. (1992). Detection of rare antigen-presenting cells by the lacZ T-cell activation assay suggests an expression cloning strategy for T-cell antigens. *Proceedings of the National Academy of Sciences of the United States of America, 89*, 6020–6024.

Lee, K. H., Holdorf, A. D., Dustin, M. L., Chan, A. C., Allen, P. M., & Shaw, A. S. (2002). T cell receptor signaling precedes immunological synapse formation. *Science, 295*, 1539–1542.

Martin-Cofreces, N. B., Vicente-Manzanares, M., & Sanchez-Madrid, F. (2018). Adhesive interactions delineate the topography of the immune synapse. *Frontiers in Cell and Development Biology, 6*, 149.

Miller, M. J., Wei, S. H., Parker, I., & Cahalan, M. D. (2002). Two-photon imaging of lymphocyte motility and antigen response in intact lymph node. *Science, 296*, 1869–1873.

Monks, C. R., Freiberg, B. A., Kupfer, H., Sciaky, N., & Kupfer, A. (1998). Three-dimensional segregation of supramolecular activation clusters in T cells. *Nature, 395*, 82–86.

Sims, T. N., Soos, T. J., Xenias, H. S., Dubin-Thaler, B., Hofman, J. M., Waite, J. C., et al. (2007). Opposing effects of PKCtheta and WASp on symmetry breaking and relocation of the immunological synapse. *Cell, 129*, 773–785.

Stinchcombe, J. C., Majorovits, E., Bossi, G., Fuller, S., & Griffiths, G. M. (2006). Centrosome polarization delivers secretory granules to the immunological synapse. *Nature, 443*, 462–465.

Stoll, S., Delon, J., Brotz, T. M., & Germain, R. N. (2002). Dynamic imaging of T cell-dendritic cell interactions in lymph nodes. *Science, 296*, 1873–1876.

Viola, A., Schroeder, S., Sakakibara, Y., & Lanzavecchia, A. (1999). T lymphocyte costimulation mediated by reorganization of membrane microdomains. *Science, 283*, 680–682.

Wabnitz, G., Kirchgessner, H., & Samstag, Y. (2019). Qualitative and quantitative analysis of the immune synapse in the human system using imaging flow cytometry. *Journal of Visualized Experiments*.

Wetzel, S. A., McKeithan, T. W., & Parker, D. C. (2002). Live-cell dynamics and the role of costimulation in immunological synapse formation. *Journal of Immunology, 169*, 6092–6101.

CHAPTER

Assessment of membrane lipid state at the natural killer cell immunological synapse

6

Yu Li and Jordan S. Orange*

Department of Pediatrics, Vagelos College of Physicians and Surgeons, Columbia University Irving Medical Center, New York, NY, United States
Corresponding author: e-mail address: jso2121@cumc.columbia.edu

Chapter outline

1 Introduction	78
2 Materials	80
2.1 Disposables	80
2.2 Equipment (parentheses state what was specifically used in our work)	80
2.3 Reagents	80
2.4 Cell lines	80
3 Common procedures	81
3.1 Preparing glass surface: Deep-cleaning	81
3.2 Preparing glass surface: Coating	81
4 Assessment of membrane lipid state via confocal microscopy	82
5 Assessment of membrane lipid state via total internal reflection fluorescence microscopy	84
6 Data analysis	86
7 Notes	87
Acknowledgments	88
References	88

Abstract

The plasma membrane is a fluid structure that protects cells as one of their first barriers and actively participates in numerous biological processes in many ways including through distinct membrane sub-regions. For immunological cells, highly organized sub-compartments of plasma membranes are vital for them to sense and react to environmental changes. This includes a varying spectrum of lipid ordering in the plasma membrane which signifies or enables cellular functions. Thus, comprehensive analyses of the plasma membrane can facilitate understanding of important cell biological elements which include insights into immune cells. Here, we describe two methods that can be used to assess membrane lipid state at the natural killer cell immunological synapse via high-resolution live cell imaging techniques.

1 Introduction

The plasma membrane is composed of a complex collection of phospholipids, sterols and membrane proteins which is dynamic in its constitution. It provides cells basic protections against external stress and regulates the exchange between the intra- and extra-cellular environments. It also actively participates in various biological processes through distinct membrane sub-regions, which are quite relevant in immunological cells (George & Wu, 2012; Taner et al., 2004).

As a heterogeneous surface, the plasma membrane has highly organized sub-compartments that are vital for an immunological cell to interact with and generate function in response to its environment. One way in which membrane sub-compartments can be created is via the ordering of the lipids themselves. Specifically, the phospholipid tails can be packed tightly owing to interspersed sterols imparting particular properties to the membrane including the ability to collect glycophosphoinositol linked proteins. Such regions of packed lipid membranes are also known as lipid rafts. These have had specific function ascribed to them in immune cells. For example, lipid rafts, can recruit, maintain and regulate multiple signaling complexes during the activation of natural killer (NK) cells (Fassett, Davis, Valter, Cohen, & Strominger, 2001; Foster, De Hoog, & Mann, 2003; Lou, Jevremovic, Billadeau, & Leibson, 2000; Sanni, Masilamani, Kabat, Coligan, & Borrego, 2004). In addition to generating a signaling platform, certain membrane domains also interact with the actin cytoskeleton to serve a role in cell motility and intracellular mechanics. NK cells and cytotoxic T lymphocytes (CTLs) can dynamically alter their membrane properties, such as packing density (Rudd-Schmidt et al., 2019), viscosity (Sheikh & Jarvis, 2011) and surface tension (Huse, 2017), to affect a variety of functions. Recently these properties have been demonstrated as important in ensuring cytotoxic lymphocyte self-survival while allowing for the elimination of target cells (Li & Orange, 2021; Rudd-Schmidt et al., 2019). Collectively, composition, distribution, interaction of membrane domains holds significant information, and their analysis can promote understanding of important cell biological questions in immunology.

Benefiting from the substantive progress in chemical biology, a wide range of fluorescent membrane probes have become commercially available and can be applied to investigate membrane properties. When combined with high-resolution live cell microscopy, these probes can enable precise observations of lipid composition, phase separation, and lipid-protein interaction, providing information on the lateral heterogeneity of and alterations in plasma membranes (Wiederschain, 2011).

Furthermore, some environmentally-sensitive probes can change their fluorescence intensities (Shvadchak, Kucherak, Afitska, Dziuba, & Yushchenko, 2017) or emission spectra (Parasassi, De Stasio, d'Ubaldo, & Gratton, 1990) in response to changes in their local environment (polarity, viscosity, packing density, etc.). These tools allow for the real-time capture of membrane dynamics, and hence greatly expand the dimensionality for increased assessment and understanding of membrane biology as it relates to cellular function.

Here, we provide two detailed protocols refined in our laboratory that can be used to assess membrane lipid state at the NK cell immunological synapse via high-resolution microscopy. In developing methods for studying the NK cell immunological synapse, we utilized live cell imaging techniques, which allows measuring biological events within optical sections of interest in living cells without interference of artifacts from fixation procedures. A model immunological synapse can be formed between an NK cell and its susceptible target carrying the ligands that trigger NK cell activation. It can be also induced between NK cells and functionalized surfaces (Rak, Mace, Banerjee, Svitkina, & Orange, 2011). We utilized two different imaging modalities, confocal and total internal reflection fluorescence (TIRF) microcopy, to analyze immunological synapses generated by these two approaches. Confocal microscopy has advantages in global comparison between the synaptic region and non-synaptic region, while TIRF microscopy can provide *en face* images of the immunological synapse with higher spatial resolution. Other imaging approaches are also relevant, but we focus on these two given how they were applied in our own laboratory. To provide a basic experimental process for discerning lipid state, we describe how to quantitatively assess lipid packing of the NK cell presynaptic membrane using the Di-4-ANEPPDHQ probe (Jin, Millard, Wuskell, Clark, & Loew, 2005). Di-4-ANEPPDHQ is a potentiometric styryl dye that shows green and red emission in the liquid-ordered and liquid-disordered phases of lipid membrane, respectively. We have used it to image phase-separated membrane domains and their dynamics. These protocols, however, are compatible with other ratiometric probes, such as Laurdan (Parasassi et al., 1990), to measure lipid packing. Given some of the characteristics of Di-4-ANEPPDHQ (Amaro, Reina, Hof, Eggeling, & Sezgin, 2017), it is important to not rely solely upon this probe in considering lipid membrane biology. The experimental approach, however, can also be applied to observe other membrane properties such as lipid composition and distribution (with suitable molecular probes) and of course in other cell types and using other imaging modalities.

The outcomes of these approaches will provide quantitative measurements of membrane order via ratiometric calculation between images taken through two distinct channels. The packing density of membrane lipid will be quantified as general polarization (GP) values. Pseudo-colored images using GP values can allow for the intuitive visualization of the distribution and dynamics of membrane microdomains of the NK cell presynaptic membrane which can help illustrate membrane biology.

2 Materials

2.1 Disposables
- Flasks and multi-well plates (polystyrene, 75 or 25 cm^2 for flasks and 6 or 12 wells for plates)
- Automatic pipette and 10 mL serological pipettes
- Micropipettes and tips (1–20 µL, 20–200 µL, 200–1000 µL)
- 8-well chamber slides (Lab-Tek II Chambered Coverglass #1.5 Sterile)

2.2 Equipment (parentheses state what was specifically used in our work)
- Cell culture incubator (Thermo Scientific 3020 Water Jacketed CO$_2$ Incubator)
- Bench top centrifuge (Eppendorf 5424R Centrifuge)
- Rotator and shaker (Thermo Scientific Tube Revolver/Rotator)
- Sonication bath (Branson M1800 Ultrasonic Cleaner with Mechanical Timer)
- Isothermal plate (Thermo Scientific Cimarec Stirring Hot Plate)
- Confocal microscope (Zeiss Axio Observer Z1 inverted microscope outfitted with a Yokogawa CSU-W1 spinning disc, with a 63×/1.4 NA objective)
- TIRF microscope (GE DeltaVision OMX-SR microscope using a 60×/1.42 PlanApoN objective)
- Image processing tools (Fiji package of ImageJ, version 1.53c)

2.3 Reagents
- Culture medium: RPMI medium 1640 (Invitrogen), 10% fetal bovine serum (FBS), 2 mM L-glutamine (Gibco), 20 mM HEPES (Gibco), Sodium pyruvate (Gibco). Adjust pH to 7.0.
- Imaging medium: Phenol red-free RPMI medium 1640 (Invitrogen), serum (FBS), 2 mM L-glutamine (Gibco), 20 mM HEPES (Gibco), Sodium pyruvate (Gibco). Adjust pH to 7.0.
- Phosphate buffered saline, without calcium and magnesium (1× PBS, Gibco)
- Poly-L-lysine solution, 0.1% (w/v) in H$_2$O (Sigma-Aldrich)
- Di-4-ANEPPDHQ (1 mM solution in DMSO, Invitrogen)
- CellMask Deep Red (1 mM solution in DMSO, Invitrogen)
- Common chemicals (Ethanol, Acetone, Sodium hydroxide)

2.4 Cell lines
- While many different types of cells can be used for this work, we have focused on using two different human NK cell lines and their corresponding classical target cells listed below.

- NK cell line YTS and corresponding target cell line 721.221 are maintained in standard culture conditions (at 37 °C, under 5% CO_2) using complete R10 medium: RPMI medium 1640 (Invitrogen), 10% fetal bovine serum (FBS), 2 mM L-glutamine (Gibco), 20 mM HEPES (Gibco), Sodium pyruvate (Gibco) adjusting pH to 7.0.
- NK cell line NK92 and corresponding target cell line K562 are maintained in standard culture conditions (at 37 °C, under 5% CO2). For the K562 cell line complete R10 medium is used. For NK92 cells Myelocult 5100 (Stemcell) with 100u/ml of IL-2 is used.

3 Common procedures
3.1 Preparing glass surface: deep-cleaning

Due to the lipophilic nature of membrane probes, a clean and fully hydrated glass surface is critical for minimizing probe non-specific binding and background noise. To prepare a high-quality glass surface for use with lipid probes:

1. Add 1 mL acetone to each well of 8-well chamber slides for 30 min (see **Note 1**).
2. Rinse chamber slides using fresh deionized water at least five times (see **Note 2**).
3. Add 2 mL NaOH solution (1 M) to each well of the 8-well chamber slides.
4. Place chamber slide on 60 °C isothermal plate for 45 min (see **Note 3**).
5. Rinse the chamber slides using fresh deionized water at least five times.
6. Rinse the chamber slides with 100% ethanol and let air dry.
7. Store the chamber slides in a clean container and use within 3 days.

3.2 Preparing glass surface: Coating

To prepare an adhesive surface for efficient cell attachment:

1. Add 0.5 mL 0.01% Poly-L-lysine solution (diluted with deionized water) at room temperature (18–26 °C) to each well of 8-well chamber slides for 1 h.
2. Rinse the chamber slides with fresh deionized water three times.
3. Dry the chamber slides in air and use them within 24 h.

To prepare an activating surface for immunological synapse formation:

1. Under room temperature (18–26 °C), add 0.25 mL antibody solution (5 µg/mL, see **Note 4**) to each well of 8-well chamber slides for 1 h.
2. Rinse the chamber slides three times using PBS.
3. Cover the glass surfaces with imaging medium and use the chamber slides in 24 h.

4 Assessment of membrane lipid state via confocal microscopy

1. Prepare clean adhesive glass surfaces (i.e., PLL coated) in chamber slides and pre-warm to 37 °C.
2. Pre-warm the imaging medium (5–10 mL) and PBS (50–100 mL) to 37 °C (see **Note 5**).
3. Pre-warm microscope stage, objectives and environmental control chamber to 37 °C (see **Note 6**).
4. Prepare NK cells (e.g., YTS cells) and target cells (e.g., 721.221 cells) by centrifuging at 300 x g, at room temperature, for 5 min.
5. Gently decant supernatants and wash cell pellets with 10 mL pre-warmed imaging medium.
6. Centrifuge both NK and target cells again at $300 \times g$, at room temperature, for 5 min.
7. Resuspend both NK and target cells with 1–3 mL pre-warmed imaging medium.
8. Add Di-4-ANEPPDHQ (recommended working concentration: 2 µM) and CellMask deep red (recommended working concentration: 1 µM) to the NK cell suspension and incubate at 37 °C, for 30 min.
9. Gently wash NK cells with 10 mL pre-warmed imaging medium.
10. Centrifuge NK cells again at $300 \times g$, at room temperature, for 5 min.
11. Repeat step 9 and 10.
12. Resuspend NK cells with 0.5 mL pre-warmed imaging medium.
13. Count both NK and target cells and adjust the concentrations of them to 0.5×10^6/ml using pre-warmed imaging medium.
14. Mix target cells and NK cells at the desired ratio (e.g., $E:T = 1:1$ or 2:1).
15. Load 200 µL of the cell suspension into the chamber wells.
16. Incubate the chamber slides at 37 °C for 20 min to allow the NK cells and target cells to form conjugates.
17. Drip immersion liquid (usually oil) on top of the lens.
18. Mount the chamber slide with cells onto the pre-warmed stage.
19. Raise the objective to touch the bottom surface of chamber slide.
20. Use transmitted light or differential interference contrast (DIC) to identify cells.
21. After finding the optimal focal plane, identify cell conjugates using differential interference contrast (DIC) or transmitted light imaging for downstream analysis.
22. Switch to the florescent channel for CellMask deep red (or other similar cell membrane dye) to confirm that a conjugate contains an NK cell (CellMask or similar positive) and target cell (CellMask or similar alternatives).
23. Adjust z-position to correspond to the z-plane crossing the center of the immunological synapse (usually indicated by approximately the largest contact region).

4 Assessment of membrane lipid state via confocal microscopy   83

24. Capture images in the DIC channel, the two ratiometric channels for Di-4-ANEPPDHQ and the CellMask deep red channel (see Note 7). 488 nm laser excites Di-4-ANEPPDHQ, and the detection ranges of the two channels are set to 500–580 nm and 620–750 nm. CellMask deep red is excited by 647 nm laser, and the detection range is set to 620–750 nm.
25. An example of this approach is provided in Fig. 1

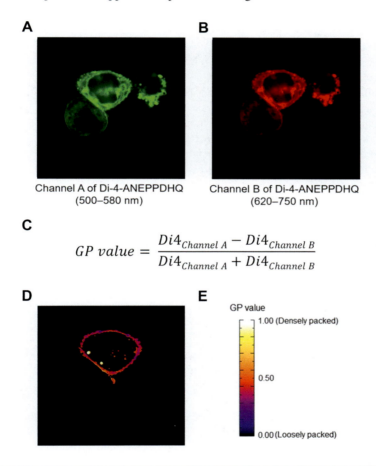

FIG. 1

The lipid packing of NK cell membranes in Di-4-ANEPPDHQ labeled YTS conjugated with 721.221 target cells, is measured using live cell confocal microscopy. Two ratiometric channels (panel A and B) for Di-4-ANEPPDHQ are captured to quantify membrane lipid density as a general polarization (GP) value. The GP value is calculated using the formula shown (panel C) and the image (panel D) is overlaid with a pseudocolor scale (panel E) to allow for the visualization of packing differences. Please note that the target cell can be seen in the Di-4-ANEPPDHQ channels A and B (see **Note 9** for additional details).

5 Assessment of membrane lipid state via total internal reflection fluorescence microscopy

1. Prepare clean activating glass surfaces (i.e., antibody coated) in chamber slides and pre-warm to 37 °C.
2. Pre-warm the imaging medium (5–10 mL) and PBS (50–100 mL) to 37 °C (see **Note 5**).
3. Pre-warm microscope stage, objectives and environmental control chamber to 37 °C (see **Note 6**).
4. Prepare NK cells (e.g., YTS cells) by centrifuging at $300 \times g$ at room temperature for 5 min.
5. Gently decant supernatants and wash cell pellets with 10 mL pre-warmed imaging medium.
6. Centrifuge cells again at $300 \times g$ at room temperature for 5 min.
7. *Re*-suspend cells with 1–3 mL pre-warmed imaging medium.
8. Add Di-4-ANEPPDHQ (recommended working concentration: 2 µM) and CellMask deep red (recommended working concentration: 1 µM) to cell suspension and incubate at 37 °C, for 30 min.
9. Gently wash cells with 10 mL pre-warmed imaging medium.
10. Centrifuge cells again at $300 \times g$, room temperature, for 5 min.
11. Repeat step 9 and 10.
12. Re-suspend cells with 0.5 mL pre-warmed imaging medium.
13. Count cells numbers and adjust the concentrations of NK cells to 0.2×10^6/mL using pre-warmed imaging medium.
14. Load 200 µL of cell suspension to chamber well.
15. Incubate chamber slides at 37 °C for 15 min to allow for immunological synapse formation.
16. Drip immersion liquid (usually oil) on top of the lens.
17. Mount the chamber slide with cells onto the pre-warmed stage.
18. Raise the objective to touch the bottom surface of chamber slide.
19. Identify fully spread NK cells using differential interference contrast (DIC) or transmitted light imaging.
20. Switch to florescent channel of CellMask deep red (or other similar cell membrane dye) to confirm cell integrity.
21. Adjust z-position and illumination depth to focus on the presynaptic membrane of NK cell and to engage total internal reflection.
22. Reduce the illumination depth or increase angle of incidence (depends on the microscope configuration, see **Note 8**), until there is a dramatic loss of fluorescence signal inside the cell. This loss of signal indicates the establishment of total internal reflection status of illumination light.
23. Increase the illumination depth or reduce the angle of incidence until the signal reappears. This is TIRF with the smallest penetration depth. As the angle of incidence is reduced, the penetration depth will increase until a certain point where total internal reflection is disrupted and TIRF transitions to epifluorescent imaging.

24. Between the maximum and minimum illumination depth or angle of incidence, choose a setting that balances the intensity and specificity of signal.
25. Capture images in DIC channel, the two ratiometric channels of Di-4-ANEPPDHQ and CellMask deep red channel Di-4- ANEPPDHQ is excited by 488 nm laser, and the detection ranges of the two channels are set to 500–580 nm and 620–750 nm. CellMask deep red is excited by 647 nm laser, and the detection range is set to 620–750 nm.
26. An example of this approach is provided in Fig. 2

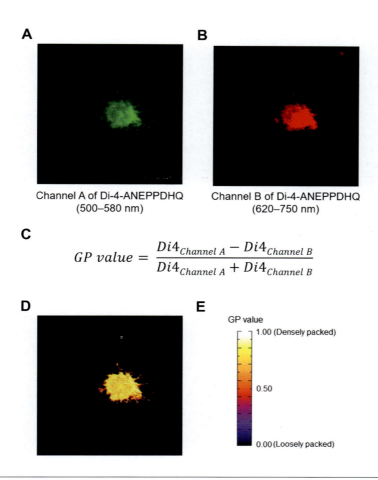

FIG. 2

The lipid packing of presynaptic membranes in Di-4-ANEPPDHQ labeled YTS is measured using TIRF microscopy. Two ratiometric channels (panel A and B) for Di-4-ANEPPDHQ are captured to quantify membrane lipid density as a general polarization (GP) value. It is calculated using the shown formula (panel C) and the images (panel D) are then overlaid with a pseudocolor scale (panel E) to allow for visualization of packing differences.

6 Data analysis

To manually quantify lipid ordering of presynaptic membranes using the Fiji version of ImageJ:

1. Download and open software (https://imagej.net/software/fiji/).
2. Import and open acquired images with Fiji (at least including three channels: two ratiometric channels for Di-4-ANEPPDHQ probe and one for CellMask deep red).
3. Generate binary masks to identify cell membrane region from images of CellMask deep red channel via the Fiji "Threshold" function (via "Image › Adjust › Threshold" or "Ctrl+Shift+T").
4. Calculate GP values from images of the two ratiometric channels via the Fiji "image calculator" function (via "Image › Image Calculator…"). The equation used to calculate the GP value is: $GP\ value = (Di4\_A - Di4\_B)/(Di4\_A + Di4\_B)$
5. Multiply the resulting images with corresponding binary masks to remove background and only preserve the GP values of the membranes of interest via the Fiji "image calculator" function.
6. To visualize GP values, choose and apply a lookup table in Fiji and set the display scale manually to "0 to 1" in the "Brightness and Contrast" window. Select a lookup table and range that is consistent with the dynamic range of the GP values to emphasize
7. To quantitatively analyze of GP values, obtain measurements via the "Analyze ›Measure" function of Fiji.
8. For time-lapse imaging series apply this procedure individually to each frame.

To perform batched imaging analysis automatically using the Fiji version of ImageJ:

1. Download and open software (https://imagej.net/software/fiji/).
2. Import and open acquired images with Fiji (at least including three channels: two ratiometric channels for Di-4-ANEPPDHQ probe and one for CellMask deep red).
3. Rename the images from 500 to 580 nm and 620–750 nm channels as "Di4_A_1, Di4_A_2, Di4_A_3, etc." and "Di4_B_1, Di4_B_2, Di4_B_3, etc.", respectively.
4. To run auto-analysis script for batch processing, paste and run the source code below via the "Process › Batch › Macro" function of Fiji. User also needs to specify the input directory in the opened dialog window as the location where all Di4 images are located.

Source code of auto-analysis script:

```
selectWindow("Di4_A.tif");
run("8-bit");
selectWindow("Di4_B.tif");
run("8-bit");
imageCalculator("Subtract create 32-bit stack", " Di4_A.tif"," Di4_B.tif");
selectWindow("Result of Di4_A.tif");
```

```
imageCalculator("Add create 32-bit", "Di4_B.tif","Di4_A.tif");
selectWindow("Result of Di4_B.tif");
imageCalculator("Divide create 32-bit", " Result of Di4_A.tif","Result of Di4_B.tif");
selectWindow("Result of Di4_A.tif");
close();
```

7 Notes

1. Only use glassware to transfer acetone. Ensure adequate ventilation via the use of a fume hood.
2. For extra cleaning measures, place the chamber slide in a beaker of deionized water (the slide should be completely immersed) and place the beaker in a sonication bath for 30 min.
3. For antibodies used to create activating surfaces we prefer anti-CD18 and anti-NKp30 to activate NK92 cells, and anti-CD18 and anti-CD28 for YTS cells. We have utilized the former for ex vivo NK cells as well.
4. Lipid rafts and the cytoskeleton can be extremely temperature sensitive from a biological standpoint. Make sure the cells are maintained at 37 °C during labeling and imaging.
5. Lipids and membranes are also temperature sensitive. Again, ensure cells are maintained at 37 °C and always use pre-warmed materials and reagents.
6. When using an immersion objective and microscope stage that have direct contact with the glass surface, be aware that heat conduction between them can greatly affect imaging quality, cell activity and membrane properties. Pre-warming the objective and stage to 37 °C can be achieved using a thermal sleeve or environmental chamber.
7. For long-term imaging like a time-lapse experiment (see also **Note 10**), it is vital to adjust laser power and exposure to minimize photo bleaching and toxicity, User is highly recommended to titrate laser power in test-runs before formal acquisition (Jonkman, Brown, Wright, Anderson, & North, 2020).
8. TIRF microscopes can have a variety different configurations and user interfaces. Please note that the illumination depth and angle of incidence are inversely correlated parameters that determine the range of fluorescence excitation.
9. Uptake of lipid dye can be seen to some degree in target cells and certainly at levels that are lower than in NK cells. The likely reasons for this are: trans-difusion, cis-diffusion and some degree of autofluorescence. For this reason it is important that lysotraker, Calcein, CellMask, or the likes used to definitively discern the NK cell in a conjugate and also allow for the generation of a binary mask for NK cell identification.

10. After NK cells establish contacts with activated surface or target cells, it takes about 8–15 min for presynaptic membrane of NK cell to accumulate high density membrane. Then it will typically maintain a stable GP value for 30–45 min, until the detachment phase of NK cell cytotoxicity.

Acknowledgments

This work was supported by NIH R01AI067946 and NYFIRST award to JSO. The authors are grateful to colleagues in their laboratory and to Dr. Emily Mace and members of her laboratory who have given feedback on ideas and experiments that have allowed for the improvement and evolution of these approaches. We are also grateful to Dr. Emily Mace for critical commentary on the written methods.

References

Amaro, M., Reina, F., Hof, M., Eggeling, C., & Sezgin, E. (2017). Laurdan and Di-4-ANEPPDHQ probe different properties of the membrane. *Journal of Physics D: Applied Physics*, *50*(13), 134004.

Fassett, M. S., Davis, D. M., Valter, M. M., Cohen, G. B., & Strominger, J. L. (2001). Signaling at the inhibitory natural killer cell immune synapse regulates lipid raft polarization but not class I MHC clustering. *Proceedings of the National Academy of Sciences of the United States of America*, *98*(25), 14547–14552.

Foster, L. J., De Hoog, C. L., & Mann, M. (2003). Unbiased quantitative proteomics of lipid rafts reveals high specificity for signaling factors. *Proceedings of the National Academy of Sciences of the United States of America*, *100*(10), 5813–5818.

George, K. S., & Wu, S. (2012). Lipid raft: A floating island of death or survival. *Toxicology and Applied Pharmacology*, *259*(3), 311–319.

Huse, M. (2017). Mechanical forces in the immune system. *Nature Reviews. Immunology*, *17*(11), 679–690.

Jin, L., Millard, A. C., Wuskell, J. P., Clark, H. A., & Loew, L. M. (2005). Cholesterol-enriched lipid domains can be visualized by di-4-ANEPPDHQ with linear and nonlinear optics. *Biophysical Journal*, *89*(1), L04–L06.

Jonkman, J., Brown, C. M., Wright, G. D., Anderson, K. I., & North, A. J. (2020). Tutorial: guidance for quantitative confocal microscopy. *Nature Protocols*, *15*(5), 1585–1611.

Li, Y., & Orange, J. S. (2021). Degranulation enhances presynaptic membrane packing, which protects NK cells from perforin-mediated autolysis. *PLoS Biology*, *19*(8), e3001328.

Lou, Z., Jevremovic, D., Billadeau, D. D., & Leibson, P. J. (2000). A balance between positive and negative signals in cytotoxic lymphocytes regulates the polarization of lipid rafts during the development of cell-mediated killing. *The Journal of Experimental Medicine*, *191*(2), 347–354.

Parasassi, T., De Stasio, G., d'Ubaldo, A., & Gratton, E. (1990). Phase fluctuation in phospholipid membranes revealed by Laurdan fluorescence. *Biophysical Journal*, *57*(6), 1179–1186.

Rak, G. D., Mace, E. M., Banerjee, P. P., Svitkina, T., & Orange, J. S. (2011). Natural killer cell lytic granule secretion occurs through a pervasive actin network at the immune synapse. *PLoS Biology, 9*(9), e1001151.

Rudd-Schmidt, J. A., Hodel, A. W., Noori, T., Lopez, J. A., Cho, H. J., Verschoor, S., et al. (2019). Lipid order and charge protect killer T cells from accidental death. *Nature Communications, 10*(1), 5396.

Sanni, T. B., Masilamani, M., Kabat, J., Coligan, J. E., & Borrego, F. (2004). Exclusion of lipid rafts and decreased mobility of CD94/NKG2A receptors at the inhibitory NK cell synapse. *Molecular Biology of the Cell, 15*(7), 3210–3223.

Sheikh, K. H., & Jarvis, S. P. (2011). Crystalline hydration structure at the membrane-fluid interface of model lipid rafts indicates a highly reactive boundary region. *Journal of the American Chemical Society, 133*(45), 18296–18303.

Shvadchak, V. V., Kucherak, O., Afitska, K., Dziuba, D., & Yushchenko, D. A. (2017). Environmentally sensitive probes for monitoring protein-membrane interactions at nanomolar concentrations. *Biochimica et Biophysica Acta - Biomembranes, 1859*(5), 852–859.

Taner, S. B., Onfelt, B., Pirinen, N. J., McCann, F. E., Magee, A. I., & Davis, D. M. (2004). Control of immune responses by trafficking cell surface proteins, vesicles and lipid rafts to and from the immunological synapse. *Traffic, 5*(9), 651–661.

Wiederschain, G. Y. (2011). The molecular probes handbook. A guide to fluorescent probes and labeling technologies. *Biochemistry (Moscow), 76*(11), 1276.

CHAPTER

Study of the effects of NK-tumor cell interaction by proteomic analysis and imaging

7

Chiara Lavarello[a], Paola Orecchia[b], Andrea Petretto[a], Massimo Vitale[b],*, Claudia Cantoni[c,d],*, and Monica Parodi[b]

[a]*Core Facilities–Clinical Proteomics and Metabolomics, IRCCS Istituto Giannina Gaslini, Genoa, Italy*
[b]*IRCCS Ospedale Policlinico San Martino Genova, Genoa, Italy*
[c]*Department of Experimental Medicine (DIMES), University of Genova, Genoa, Italy*
[d]*Laboratory of Clinical and Experimental Immunology, IRCCS Istituto Giannina Gaslini, Genoa, Italy*
**Corresponding authors: e-mail address:* massimo.vitale@hsanmartino.it; claudia.cantoni@unige.it

Chapter outline

1 Introduction...93
2 Melanoma/NK cell co-culture...93
　2.1 Materials...94
　　2.1.1 Common disposables..94
　　2.1.2 Equipment...94
　　2.1.3 Reagents...94
　2.2 Methods..95
　　2.2.1 Direct isolation of untouched NK cells from whole blood............95
　　2.2.2 Generation of polyclonal NK cell lines.......................................95
　　2.2.3 Melanoma Cell lines and check of their EMT status..................96
　　2.2.4 Melanoma/NK cell co-culture and EMT induction.....................96
　　2.2.5 Melanoma/NK cell co-culture for cell imaging...........................97
3 Cell imaging...97
　3.1 Materials...98
　　3.1.1 Common disposables..98
　　3.1.2 Equipment...98

      3.1.3 Reagents..98
  3.2 Methods...98
      3.2.1 Fixation and permeabilization..98
      3.2.2 Indirect immunofluorescence..99
      3.2.3 Direct immunofluorescence...99
      3.2.4 Nuclei counterstaining...99
4 Analysis of the proteomic changes related to the process of EMT induced by NK cells..99
  4.1 Materials...101
      4.1.1 Common disposables..101
      4.1.2 Equipment...101
      4.1.3 Reagents...101
  4.2 Methods...102
      4.2.1 Sample preparation..102
      4.2.2 Protein concentration determination..............................102
      4.2.3 Mass spectrometer analysis...103
      4.2.4 Data and statistical analyses..104
Notes..105
Acknowledgments..106
Declaration of interests..107
References..107

Abstract

Natural Killer (NK) cells play a pivotal role in the elimination of tumor cells. The interactions that NK cells can establish with cancer cells in the tumor microenvironment (TME) are crucial for the outcome of the anti-tumor response, possibly resulting in mechanisms able to modulate NK cell effector functions on the one side, and to modify tumor cell phenotype and properties on the other side.

    This chapter will describe two different experimental approaches for the evaluation of NK-tumor cell interactions. First, a detailed protocol for the setting up of NK-tumor cell co-cultures will be illustrated, followed by information on cell imaging techniques, useful for assessing cell morphology and cytoskeletal changes. The second part will be focused on the description of a proteomic approach aimed at investigating the effect of this crosstalk from another point of view, i.e., characterizing the cellular and molecular pathways modulated in tumor cells following interaction with NK cells.

    The chapter centers on the interaction between NK and melanoma cells and refers to experimental approaches we set up to study the effects of this cross-talk on the process of the Epithelial-to-Mesenchymal Transition (EMT). Nevertheless, the described protocols can be quite easily adapted to study the interaction of NK cells with adherent tumor cell lines of different origin and histotype, as in our original study, we also analyzed possible NK-induced morphologic changes in the cervix adenocarcinoma HeLa cells and the colon cancer HT29 cells.

1 Introduction

Natural Killer (NK) cells are mainly known for their cytotoxic activity, which marks them as powerful effectors of innate immunity involved in the control of viral infection and tumors. However, NK cells have been increasingly appreciated also for their phenotypic and functional heterogeneity and for their regulatory properties, related to their capability of releasing soluble factors, including cytokines and chemokines (Sivori et al., 2019; Vitale et al., 2019). Therefore, in the context of the NK cell-mediated anti-tumor response, the study of possible NK-tumor cell cross-talk becomes crucial. Specifically, the characterization of this interaction is relevant to a deeper understanding of the local tumor microenvironment and, possibly, to the design of effective NK-based immunotherapy strategies. Several studies have shown and characterized tumor cell-mediated inhibitory mechanisms targeting NK cell function, spanning from the shedding of soluble NK-receptor ligands to the release of immune-suppressive factors, and to the impairment of the immunological synapse. On the NK cell side, NK cells have been shown to influence tumor cell phenotype and properties.

In this regard, we have recently described the effects of NK cells on melanoma cells, specifically on their epithelial-mesenchymal transition (EMT) (Huergo-Zapico et al., 2018). The EMT process typically involves changes in transcription factor expression leading to the acquisition of morphologic and functional properties associated with increased invasiveness, drug resistance, and metastatic spread.

In our study we set a series of co-culture experiments and utilized different experimental approaches and assays to dissect the functional interactions and define the underlying mechanisms. In particular, we evaluated the effects that NK cells exerted on tumor cell properties by the use of imaging techniques, aimed at assessing tumor cell morphology and cytoskeletal changes, as well as of proteomic analysis, aimed at revealing modulated molecular and cellular pathways.

In this chapter we will describe methodologies utilized: (1) to prepare co-culture experiments, (2) to assess morphology and cytoskeletal changes by cell imaging, and (3) to define proteomic changes related to the process of EMT possibly induced by NK cells.

Remarkably, the co-culture protocols and the described analytical evaluations can be applied to a broader type of studies, aimed at defining the influence of NK cells on different aspects of tumor cell biology, including remodeling of tumor tissues and vascularization, generation of circulating tumor spheroids, or acquisition of proteomic profiles related to tumor progression and metastatic spread. In particular, thanks to ever more effective preparative and analytic approaches, including Single Cell analysis, proteomics can be crucial to the precise characterization of cell subsets and diseases (see also Note 16).

2 Melanoma/NK cell co-culture

The assessment of the effects that NK cells exert on melanoma cells, can be evaluated by setting co-culture experiments using peripheral blood (PB)-derived NK cells (in vitro expanded) and melanoma cell lines obtained from metastatic melanoma specimens. Specifically assessing EMT, melanoma cell cultures (without NK cells) should be exposed to appropriate EMT-inducing cytokines for comparison.

2.1 Materials
2.1.1 Common disposables
- Serological pipettes (5 and 10 mL)
- Polystyrene flasks (75 cm^2)
- U-bottom-96-well plates
- 6-well cell culture plates
- 50 mL, 14 mL polypropylene conical base tubes
- Polystyrene round bottom tubes 12 × 75 mm
- Neubauer cell count chamber
- Forceps
- 22 × 22 mm glass coverslip for 6-well plates
- 26 × 76 mm microscope slides

2.1.2 Equipment
- Sterile cell culture laminar flow hood (safety level II)
- Cell culture incubator for standard cell culture conditions in a humidified atmosphere (37 °C, 5% CO_2)
- Centrifuge
- Micropipettes (0.2–2 μL, 1–20 μL, 20–200 μL, 200–1000 μL) and multichannel micropipettes (50–300 μL) and tips
- Autoclave
- Cell Irradiator
- Flow cytometry instrumentation

2.1.3 Reagents
- Cell culture medium: RPMI1640
- Complete medium: RPMI1640 supplemented with 10% heat-inactivated Foetal Bovine Serum (FBS), 2 mM L-glutamine, 100 U/mL Potassium Penicillin 100 μg/mL Streptomycin Sulphate (Pen-Strep)
- Complete medium + IL-2 (i.e., supplemented with 100 IU/mL human recombinant IL-2)
- Phosphate buffer saline (PBS)
- 1.0770 g/cm^3 density gradient medium Lympholyte-H
- Trypan Blue
- Trypsin-EDTA 1× solution
- Phytohemagglutinin (PHA)
- RosetteSep Human NK enrichment cocktail (Stem Cell Technologies 15065)
- Primary antibodies and fluorochrome-conjugated specific antibodies

2.2 Methods

2.2.1 Direct isolation of untouched NK cells from whole blood

Natural Killer (NK) cells can be directly isolated from whole blood by negative selection using the RosetteSep™ Human NK Cell Enrichment Cocktail. Rosette-Sep™ Tetrameric Antibody Complexes, recognizing non-NK cells and red blood cells (RBCs), target unwanted cells to RBCs to form "rosettes." After density gradient centrifugation, unwanted cells pellet along with RBC and purified NK cells are present at the interface between the plasma and the gradient density medium.

- Transfer the anticoagulated whole blood (5–10 mL) in a 50 mL tube.
- Add RosetteSep Human NK Cell Enrichment Cocktail (50 µL/mL of whole blood).
- Incubate 20 min at room temperature mixing gently.
- Dilute the sample with an equal volume of RPMI1640+2% FBS, mix gently and lay on Lympholyte-H cell separation medium. Centrifuge 20 min at room temperature at $1200 \times g$ (brake off).
- Collect enriched NK cells from the Lympholyte-H/plasma interface and wash twice in complete medium.
- Evaluate NK cell number and purity (see Note 1).

2.2.2 Generation of polyclonal NK cell lines

Polyclonal NK cell lines can be obtained by culturing purified NK cells on feeder cells (irradiated Peripheral Blood Mononuclear Cells, PBMCs) in the presence of 100 U/mL rIL-2 and 1.5 ng/mL PHA in round-bottomed 96-well microtiter plates. After 15–20 days of culture, the expanded NK cells are ready to be used.

- Transfer the anticoagulated whole blood (5–10 mL) in a 50 mL tube.
- Dilute the sample with an equal volume of RPMI1640+2% FBS, mix gently and layer on Lympholyte-H cell separation medium. Centrifuge 20 min at room temperature at $1000 \times g$ (brake off).
- Collect PBMCs from the Lympholyte-H/plasma interface and wash twice in complete medium.
- Resuspend PBMCs at a density of $2–5 \times 10^6$ cells/mL in complete medium.
- Irradiate with 5000 rad.
- Centrifuge cells at 4°C for 7 min at $300 \times g$.
- Discard the supernatant.
- Add 10 mL of complete medium.
- Carefully count cells by trypan blue exclusion method using Neubauer cell count chamber.
- Discard the supernatant and resuspend irradiated PBMCs (feeder) at a density of 1.5×10^6 cells/mL in complete medium + IL-2 containing 1.5 ng/mL PHA.
- Plate in 96-well U-bottom plates, 100 µL of the suspension to each well.
- Resuspend NK cells (see above) in complete medium + IL-2 at a density of $10^4–10^5$ cells/mL.
- Plate in 96-well U-bottom plates, 100 µL of the suspension in the wells containing irradiated feeder cells.

The expanded NK cells can be used in co-culture experiments.

2.2.3 Melanoma Cell lines and check of their EMT status

In the original study, melanoma cell lines were derived from metastatic melanoma resections. Once established, the primary cell lines were expanded for few passages and phenotypically characterized by flow cytometry. Cells were thawed, tested for mycoplasma by Polymerase Chain Reaction (PCR) and used within 30 days. The melanoma cell cultures were phenotypically characterized and assessed for their purity by the analysis of informative surface markers including Mel-CAM/CD146 and GD2. To assess their initial EMT status, melanoma cells were also analyzed for the expression of E-cadherin (highly expressed on epithelial cells) and of N-cadherin and fibronectin (typical mesenchymal markers). Alternatively, commercially available melanoma cell lines can be used. Also in this case, their EMT status should be checked as above.

2.2.4 Melanoma/NK cell co-culture and EMT induction

- Melanoma cells are normally cultured in 75 cm^2 cell culture flasks.
- Remove the medium from the flask and wash twice with 10 mL of PBS to eliminate FBS (which would limit trypsin activity).
- Add 2 mL of trypsin-EDTA 1× solution.
- Place for 10 min in a CO_2 incubator at 37 °C.
- Check cell detachment at the microscope and shake the flask to completely detach residual adherent cells.
- Add 8 mL of complete medium (proteins contained in FBS will saturate trypsin and block its activity).
- Remove from the flask the supernatant containing cells and transfer in 14 mL tube.
- Centrifuge cells at 4 °C for 7 min at 300 × g.
- Discard the supernatant and resuspend in 10 mL of complete medium.
- Carefully count cells by trypan blue exclusion method using Neubauer cell count chamber.
- Centrifuge cells at 4 °C for 7 min at 300 × g.
- Resuspend melanoma cells at a density of $0.75–1 \times 10^5$ cells/mL (see Note 2) in complete medium.
- In a 6-well cell culture plate place 2 mL of melanoma cell suspension/well and transfer in a CO_2 incubator.
- After 16 h, collect NK cells from polyclonal NK cell cultures, dilute the cells 1:2 with complete culture medium and centrifuge at 300 × g for 7 min.
- Eliminate supernatant and resuspend in fresh complete medium + IL-2 up to the original volume.
- Carefully count cells by trypan blue exclusion method using Neubauer cell count chamber.
- Adjust NK cell concentration using complete medium + IL-2 at $4–5 \times 10^5$ cells/mL (see Note 3).
- Wash melanoma cells after 16-h culture in 6-well plate: remove culture medium and add 2 mL of PBS to each well.

3 Cell imaging

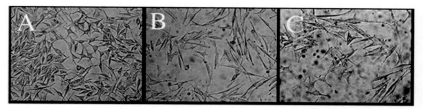

FIG. 1

Morphology analysis of live melanoma cells undergoing EMT. Cells undergoing EMT acquire a spindle shaped morphology, which can be observed directly on live cells using an inverted microscope. Melanoma cells (from the MeCop melanoma cell line) have been cultured for 96 h (A) in complete medium +100 IU IL-2 (as control) without NK cells, (B) in complete medium + EMT-inducing cytokines (TGF-β and TNF-α) (as positive control), (C) in complete medium +100 IU IL-2 + NK cells.

- Remove PBS and add complete medium + IL-2 (as control culture) (see Note 4), or complete medium + IL-2 and 100 μL of polyclonal NK cell suspension, or complete medium containing 5 ng/mL TGF-β1 + 10 ng/mL TNF-α (to induce maximal EMT—i.e., positive control).
- Place the plate in a CO_2 incubator for 96 h.

After 96 h, cells can be checked for EMT transition by flow cytometry (see Note 5) or by analyzing their morphology (using an inverted microscope) (Fig. 1) (see Note 6).

2.2.5 Melanoma/NK cell co-culture for cell imaging

Analysis is done on fixed cells and methods of co-culture preparation are modified as follows:

- Using a sterile pair of forceps, place sterile glass coverslips in a 6-well cell culture plate (see Note 7).
- Place adherent melanoma cells on glass coverslips and culture them as indicated above (see Section 2.2.4). Ensure that the cell suspension is actually applied on the coverslip by gentle pressing on the coverslip with a tip avoiding growth of cells under the coverslip.

3 Cell imaging

In order to evaluate EMT-related changes in melanoma cells, it is possible to analyze cadherin switch (i.e., decreased expression of E-cadherin and increased N-Cadherin), fibronectin expression and F-actin filament reorganization. These analyses are done on fixed cells. The co-culture methods to obtain cells for fixation on coverslips are described in Section 2.2.5.

The cell imaging techniques described here, were previously used for the analysis of different biological processes related to tumor biology (Huergo-Zapico et al., 2018; Orecchia et al., 2015, 2019).

3.1 Materials
3.1.1 Common disposables
- 22 × 22 mm glass coverslips for 6-well plates
- 26 × 76 mm microscope slides
- Forceps

3.1.2 Equipment
- Centrifuge
- Micropipettes (0.2–2 μL, 1–20 μL, 20–200 μL, 200–1000 μL) and tips
- Fume hood
- Autoclave
- Vortex
- Light optical microscope
- Fluorescence and confocal microscopes

3.1.3 Reagents
- Phosphate buffer saline (PBS)
- 4% Paraformaldehyde (PFA) (See Note 8)
- Triton-X100
- 100% methanol (MeOH) (See Note 8)
- Bovine Serum Albumin (BSA)
- DAPI
- Hoechst dye
- Mounting medium with or without DAPI

3.2 Methods
3.2.1 Fixation and permeabilization
- All the following steps are performed directly in the plate, where co-cultures were performed on coverslip (see Section 2.2.5).
- Remove the culture medium and rinse the coverslips twice in PBS.
- Fix coverslips in 4% PFA/PBS for 30 min at room temperature (RT). PFA cross-links proteins so that they do not solubilize when cells are permeabilized.
- After fixation, remove PFA and permeabilize cells in PBS/0.1% Triton X-100 for 10 min at RT and allow to air dry for 10 min. Otherwise, fix in cold MeOH for 10 min. MeOH works by precipitating proteins and allows permeabilization without detergents (see Note 9).

3.2.2 Indirect immunofluorescence
- Incubate the coverslips for 30 min at RT in 1% BSA/PBS to block nonspecific binding sites.
- After the blocking step, drain the coverslip.
- Incubate the coverslip in the primary antibody solution for 45 min at RT in the dark. To prepare this solution, dilute the primary antibody with 1% BSA in PBS or in "antibody" buffer (see Note 10). Negative controls with irrelevant or no primary antibody must be set up for every staining experiment. In this setting, primary antibodies include: anti-E-cadherin (67A4) and anti-N-cadherin (8C11) (Santa Cruz Biotechnology, Dallas, USA), anti-Fibronectin (IgG1, IST-4).
- Rinse the coverslip twice in PBS (see Note 11).
- Incubate the coverslip with the fluorochrome-conjugated isotype-specific secondary antibody diluted with 1% BSA in PBS (according to the manufacturer's instructions) for 30 min at RT in the dark. For imaging, it is suggested to use bright, photostable fluorochromes not sensitive to pH. If immunostaining of the cell sample with a second primary antibody is required, sequential staining of each primary antibody is recommended. The two primary antibodies usually should be from different species.

3.2.3 Direct immunofluorescence
- Incubate the coverslip with the fluorochrome-conjugated primary antibody diluted in 1% BSA/PBS for 30 min in the dark at RT (see Note 10).
- Rinse the coverslip twice in PBS.

3.2.4 Nuclei counterstaining
- Incubate the coverslip with a counterstain solution to stain nuclei, like Hoechst dye or DAPI according to the manufacturer's instructions.
- Rinse the coverslip twice in PBS.
- Using forceps, place the coverslip (keep cell side face down) onto a drop of aqueous, anti-fade mounting medium on a 26 × 76 mm glass microscope avoiding air bubbles (see Note 12).
- Allow to air dry overnight in the dark and clean around the edges before examining under the microscope.
- Slides can be stored for several months in the dark at 4 °C.
- Images are captured using fluorescence or confocal microscopy.

4 Analysis of the proteomic changes related to the process of EMT induced by NK cells

Melanoma cells cultured alone, co-cultured with NK cells, or exposed to EMT-inducing cytokines can be analyzed by a proteomic approach in order to assess possible signatures. The complete experimental workflow is illustrated in Fig. 2.

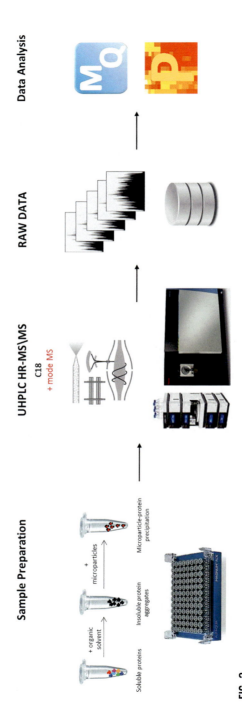

FIG. 2

Complete workflow for protein discovery analysis. It includes sample preparation through immobilization on microparticles by protein aggregation capture (Kulak, Pichler, Paron, Nagaraj, & Mann, 2014), data-dependent analysis on high resolution mass spectrometry, and data and statistical analyses.

4.1 Materials
4.1.1 Common disposables
- Micropipettes (0.2–2 µL, 1–20 µL, 20–200 µL, 200–1000 µL) and tips
- Eppendorf LoBind microcentrifuge tubes (Catalog number 0022431021)
- Eppendorf microcentrifuge tubes
- Eppendorf epT.I.P.S. (Catalog number 0030073266)
- Polystyrene-divinylbenzene copolymer partially modified with sulfonic acid (SDB-RPS) disks (3 M Catalog number 2241)
- Costar 96-well plates (catalog number 3915)
- Nano RP chromatographic column Easy Spray (ES803, Thermo Scientific)

4.1.2 Equipment
- Vortex
- Speed vac
- Thermomixer
- Centrifuge
- Fluorescence spectrophotometer (Infinite 200, Tecan)
- High resolution mass spectrometer (MS) (Orbitrap Tribrid Fusion, Thermo Scienfific)

4.1.3 Reagents
- Acetonitrile (ACN) LC-MS grade
- Water LC-MS grade
- Formic acid (FA)
- Trifluoroacetic Acid (TFA)
- 10 mM Tris(2-carboxyethyl)phosphine (TCEP) hydrochloride
- 6 M Guanidine hydrochloride (GdmCl)
- 40 mM Chloroacetamide (CAA)
- 100 mM Tris (Tris(hydroxymethyl)aminomethane) pH 8.5
- ProteaseMAX™ Surfactant (Catalog number:V2071, Promega)
- Trypsin enzyme
- LysC enzyme
- Ammonium hydroxide
- Dimethyl sulfoxide (DMSO)
- Ethanol (EtOH)
- Isopropanol
- 8 M Urea
- 0.1 µg/µL Tryptophan
- SpeedBead Magnetic Carboxylate (Catalog number: 45152105050250 and 65,152,105,050,250)
- Pierce™ LTQ Velos ESI Positive Ion Calibration Solution (Catalog number: 88323)

4.2 Methods

4.2.1 Sample preparation

In the original study, samples were processed by in-Stage Tip protocol (Kulak et al., 2014) with a little modification of lysis solvent.

- Lyse cells in 6 M GdmCl, 10 mM TCEP, 40 mM CAA, 100 mM Tris pH 8.5 buffer by alternating between heating (99 °C) and freezing (−20 °C) cycles and sonicate in the ultrasonic bath (see Note 13).
- Determine protein concentration by tryptophan fluorescence (see Section 4.2.2).
- Load 20 μg of proteins in three 14-gauge SDB-RPS home-made tip plugs and add 0.5% ProteaseMAX™ Surfactant (see Note 14).
- Dilute 1:10 with 10% (v/v) ACN, 25 mM Tris pH 8.5 containing 0.5 μg of trypsin enzyme (the ratio micrograms of enzyme to micrograms of protein should be 1:50).
- Digest overnight at 37 °C shaking at 400 rpm in the thermomixer.
- To stop the digestion reaction, add 0.5% (v/v) TFA (check the pH, it should be less then 3).
- Centrifuge the tips at a maximum speed of 2000 × g.
- Wash the tips three times with 100 μL 0.2% (v/v) TFA.
- Elute the peptide in a new Eppendorf tube with 5% (v/v) ammonium hydroxide, 80% (v/v) ACN by centrifugation at 1400 × g.
- Dry the sample with a speed vacuum system and then resuspend it in 25 μL of 2% ACN solution in H_2O with 0,1% FA.

Alternatively, samples can be prepared with a more recent protocol, termed PAC or SP3, which exploits the phenomenon of nonspecifically immobilizing precipitated and aggregated proteins on any type of sub-micron particles irrespective of their surface chemistry (see Notes 15 and 16).

4.2.2 Protein concentration determination

Protein concentrations can be determined by tryptophan fluorescence emission at 350 nm using an excitation wavelength of 295 nm. The emission at 350 nm of 0.1 μg of tryptophan is equivalent to 7 μg of protein on average (see Note 17).

- Prepare a stock solution of tryptophan at 10 mg/mL in 50% ACN.
- Dilute the stock solution to obtain a working solution = 0.1 μg/μL in 8 M Urea, aliquot into Eppendorf tubes, and store at −80 °C.
- In a 96-well plate add in each well 200 μL of 8 M Urea and use the stock solution to build a standard calibration curve (0.25–1.5 μL).
- In a 96-well plate add also the samples, acquire the fluorescence emission, and calculate the protein concentration of the sample using the calibration curve equation.

4.2.3 Mass spectrometer analysis

- Perform LC-MS/MS (tandem mass spectrometry) measurements by coupling an RS 3000 ultimate HPLC (high-performance liquid chromatography) with Orbitrap Tribrid Fusion instrument.
- Directly inject 5 μL of sample in the nano RP (reversed-phase) column and separate the peptides at a flow rate of 250 nL/min using a 180 min gradient from 0 to 45% solvent B (0–3 min: 2% B; 3–7 min: 2–5% B; 7–140 min 5–30% B, 140–165 min: 30–45% B, 165–165.1 min: 45–80% B, 165.1–170 min: 80% B, 170–180 min: 80–2% B).
- A and B eluents are respectively 98/2 (*v/v*) H20/ACN with 0.1% FA and 80/20 (v/v) ACN/H$_2$0 with 5% DMSO and 0.1% FA (Hahne et al., 2013).
- During peptide elution, directly inject peptides into the mass spectrometer via electrospray ionization in positive ionization mode, and operate the Orbitrap Tribrid Fusion in data-dependent mode, automatically switching between MS1 and MS2.
- Acquire full-scan MS spectra at 375–1500 m/z, 120,000 resolution with acquisition gain control (AGC) target value of 2.5×10^5 charges and maximum injection time of 50 ms for MS1. Acquire MS2 spectra at 17500 resolution, AGC target value of 1×10^4 charges and max injection time of 45 msec.

Suggested parameters for data-dependent acquisition on a Tribrid Fusion are specified in Table 1 (see Note 18).

Table 1 Mass spectrometer parameters.

Full MS	
Resolution	120,000
AGC target	250,000
Maximum IT	50 ms
Scan range	360–1300 m/z
MS2	
Resolution	17,500
AGC target	1e4
Maximum IT	45 ms
Cycle time	2 s
Isolation mode	quadrupole
Isolation window	1.4 m/z
Activation type	HCD
Collision energy	30
Include charge states	2–5

4.2.4 Data and statistical analyses
- Analyze the data using a proteomics software capable of performing label-free quantification. Results can be obtained based on peptide identifications by the search of raw data against the UniProtKB human database, using the freely available MaxQuant (Tyanova, Temu, & Cox, 2016) and Andromeda search engine (Cox et al., 2011).

Parameters applied are specified in Table 2.

- Use Perseus software (Tyanova et al., 2016) to analyze the MaxQuant output.
- To obtain the number of protein groups and unique genes open the proteinGroups.txt and exclude reverse and contaminant hits.
- Filter the protein group to require a reasonable percentage of valid values in at least one experimental group (usually a percentage between 70 and 100% is chosen).
- Transform the data to a logarithmic scale (e.g., log2(x)) and, only if it has filtered out proteins with a percentage other than 100%, imputes to NaN (Not a Number) values random numbers from a normal distribution separately for each column.

Table 2 MaxQuant parameters.

Group-specific parameters	
Type	Standard
Label	No
Variable modifications	Acetyl (Protein N-term), oxidation (M), deamidation (Q)
Digestion mode	Specific (Trypsin/P)
Max. missed cleavages	2
Main search peptide tolerance	4.5 ppm
Max. number of modifications per peptide	5
Global parameters	
Database	UniProtKB
Fixed modifications	Cysteine carbamidomethylation
PSM FDR	0.01
Protein FDR	0.05
Site decoy fraction	0.01
Min. peptide length	7
Min. score for unmodified peptides	0
Min. score for modified peptides	40
MS/MS match tolerance	20 ppm
Min. delta score for modified peptides	6

PSM: peptide spectrum match; FDR: false discovery rate.

- To identify proteins that show differential expression between two groups apply a permutation-based t-test (FDR=0.05 and S0=0.1).
- To visualize the profile of the significative proteins, normalize data using Z-score and perform hierarchical clustering using Euclidean distances.

Notes

1. Use flow cytometry to verify the purity of NK cell population (95% of CD56$^+$ CD3$^-$ CD14$^-$).
2. The cell concentration should be set in order to have 75% confluence when NK cells will be added to the culture (i.e., after overnight tumor cell culture). The seeding concentration may vary depending on the proliferation rate of the different melanoma cell lines (or other adherent tumor cells), therefore some preliminary setting is recommended.
3. The NK cell concentration may be further adjusted depending on the susceptibility of the different melanoma cell lines (or other adherent tumor cells) to NK cell-mediated lysis.
4. IL-2 has no effect on the EMT process in melanoma cells. This cytokine is added in co-culture to sustain vitality and function of NK cells, and in control culture to appropriately evaluate the effects of the sole NK cells. For other tumor cell lines, check the effect of IL-2 prior to proceed with the co-culture experiments.
5. Evaluation of EMT by flow cytometry is done analyzing E-Cadherin (whose expression generally decreases upon EMT), and N-Cadherin and Fibronectin (increased expression after EMT).
6. Cells undergoing EMT acquire a spindle shaped morphology, which can be directly observed on live cells using an inverted microscope (Fig. 1). This morphologic change is not clearly visible in all melanoma cells undergoing EMT. In this latter case, the EMT process is marked by the Cadherin switch that can be assessed by FACS.
7. Before seeding cells, autoclave glass coverslips on the dry cycle for 20 min.
8. Handle only under a fume hood and wear gloves.
9. Plates can be stored for several months at −20 °C after sealing them with parafilm to prevent drying during storage.
10. The antibody optimal dilution should be determined in a separate experiment on the same cells by titration curve.
11. All washing steps could be performed in 2% sucrose/PBS to stabilize the cell membrane.
12. Aqueous mounting medium with DAPI could be used to avoid nuclear counterstaining step.
13. The combined reduction and alkylation in one step is possible thanks to the chemical compatibility of TCEP-HCl (hydrochloric acid) and CAA, this saves time in the sample preparation step.

14. ProteaseMAX™ Surfactant is acid-labile and should be dissolved in freshly prepared ammonium bicarbonate buffer (pH ~7.8). Dissolving the surfactant in a buffer of lower pH will degrade the surfactant. Using ProteaseMAX™ Surfactant at concentrations higher than those suggested may lead to loss of peptide signal due to precipitation of the peptides.
15. Here we report protein aggregation capture (PAC) method (Batth et al., 2019).
    - Add the 1:1 mixture of beads to the sample in a ratio of at least 1:4 sample: beads (1 μL = 50 μg beads).
    - Add ACN to the sample to get a 70% solution.
    - Vortex for at least 30 s and let the beads precipitate for 10 min (repeat twice).
    - Transfer to magnet, wait at least 1 min and discard the supernatant.
    - Without removing the samples from the magnet wash with a mix 1:1:1 ACN, 70% EtOH, isopropanol (repeat 3 times). Before removing the magnet wait for the beads to be re-deposited (at least 10 s).
    - Allow the beads to dry from the isopropanol and add 100 μL Tris 25 mM pH 8.
    - Digest the proteins with trypsin and LysC overnight at 37°C.
    - Transfer to magnet and recover supernatants that will be acidified with 10 μL TFA 2%.
16. Novel biochemical approaches, in combination with recent developments in mass spectrometry-based proteomics instrumentation and data analysis pipelines, have now enabled the dissection of disease phenotypes and their modulation at unprecedented resolution and dimensionality (Meissner, Geddes-McAlister, Mann, & Bantscheff, 2022). Moreover, Single Cell Proteomics, based on multiplex or label-free approach, is the clearest demonstration of how today it is possible to overcome problems of low concentrations even in complex and heterogeneous samples (Kelly, 2020). The sample preparation methods mentioned, both the newer one called PAC and the in-Stage Tip method, are considered universal for label-free quantitative proteomic analysis samples. Therefore, by applying these sample preparation methods and studying a proper experimental design, the sensitivity problem of mass spectrometry on biological samples in shotgun proteomics is almost irrelevant.
17. This method is indicated only for cell lysates because it is based on an average presence of tryptophan in the sample.
18. Since each LC-MS instrument has different characteristics such as dead volume, the cycle time has to be adjusted to the chromatographic peak width.

Acknowledgments

Fondazione AIRC, grant number IG 2020 id. 25023 (M.V.); Italian Ministry of Health (project RF-2018-12366714 (M.V.); Italian Ministry of Health (Ricerca Corrente 2022-24 and 5x1000 2018-19).

Declaration of interests

The Authors declare no conflicts of interest.

References

Batth, T. S., Tollenaere, M. X., Rüther, P., Gonzalez-Franquesa, A., Prabhakar, B. S., Bekker-Jensen, S., et al. (2019). Protein aggregation capture on microparticles enables multipurpose proteomics sample preparation. *Molecular & Cellular Proteomics*, *18*(5), 1027–1035.

Cox, J., Neuhauser, N., Michalski, A., Scheltema, R. A., Olsen, J. V., & Mann, M. (2011). Andromeda: A peptide search engine integrated into the MaxQuant environment. *Journal of Proteome Research*, *10*(4), 1794–1805.

Hahne, H., Pachl, F., Ruprecht, B., Maier, S. K., Klaeger, S., Helm, D., et al. (2013). DMSO enhances electrospray response, boosting sensitivity of proteomic experiments. *Nature Methods*, *10*, 989–991.

Huergo-Zapico, L., Parodi, M., Cantoni, C., Lavarello, C., Fernández-Martínez, J. L., Petretto, A., et al. (2018). NK-cell editing mediates epithelial-to-mesenchymal transition via phenotypic and proteomic changes in melanoma cell lines. *Cancer Research*, *78*(14), 3913–3925.

Kelly, R. T. (2020). Single-cell proteomics: Progress and prospects. *Molecular & Cellular Proteomics*, *19*(11), 1739–1748.

Kulak, N. A., Pichler, G., Paron, I., Nagaraj, N., & Mann, M. (2014). Minimal, encapsulated proteomic-sample processing applied to copy-number estimation in eukaryotic cells. *Nature Methods*, *11*, 319–324.

Meissner, F., Geddes-McAlister, J., Mann, M., & Bantscheff, M. (2022). The emerging role of mass spectrometry-based proteomics in drug discovery. *Nature Reviews. Drug Discovery*.

Orecchia, P., Balza, E., Pietra, G., Conte, R., Bizzarri, N., Ferrero, S., et al. (2019). L19-IL2 Immunocytokine in combination with the anti-Syndecan-1 46F2SIP antibody format: A new targeted treatment approach in an ovarian carcinoma model. *Cancers (Basel)*, *11*(9), 1232.

Orecchia, P., Conte, R., Balza, E., Pietra, G., Mingari, M. C., & Carnemolla, B. (2015). Targeting Syndecan-1, a molecule implicated in the process of vasculogenic mimicry, enhances the therapeutic efficacy of the L19-IL2 immunocytokine in human melanoma xenografts. *Oncotarget*, *6*(35), 37426–37442.

Sivori, S., Meazza, R., Quintarelli, C., Carlomagno, S., Della Chiesa, M., Falco, M., et al. (2019). NK cell-based immunotherapy for hematological malignancies. *Journal of Clinical Medicine*, *8*, 1702.

Tyanova, S., Temu, T., & Cox, J. (2016). The MaxQuant computational platform for mass spectrometry-based shotgun proteomics. *Nature Protocols*, *11*, 2301–2319.

Tyanova, S., Temu, T., Sinitcyn, P., Carlson, A., Hein, M. Y., Geiger, T., et al. (2016). The Perseus computational platform for comprehensive analysis of (prote)omics data. *Nature Methods*, *13*, 731–740.

Vitale, M., Cantoni, C., Della Chiesa, M., Ferlazzo, G., Carlomagno, S., Pende, D., et al. (2019). An historical overview: The discovery of how NK cells can kill enemies, recruit defense troops, and more. *Frontiers in Immunology*, *10*, 1415.

CHAPTER 8

Protocol for the murine antibody-dependent cellular phagocytosis assay

Eliana Stanganello, Magdalena Brkic, Steven Zenner, Ines Beulshausen, Ute Schmitt, and Fulvia Vascotto*

TRON - Translational Oncology at the University Medical Center of the Johannes Gutenberg University GmbH, Mainz, Germany
*Corresponding author: e-mail address: fulvia.vascotto@tron-mainz.de

Chapter outline

1 Introduction	110
2 Materials	112
2.1 Disposable materials	112
2.2 Reagents	112
2.3 Equipments	113
2.4 Cells and mice	113
2.5 Cell culture	114
2.6 Software	114
2.7 Methods	114
2.7.1 Cell culture	114
2.7.2 MC38-Her2+ electroporation	114
2.7.3 ADCP assay	115
2.7.4 Flow cytometry	116
2.7.5 Immunofluorescence and confocal imaging	116
3 Notes	118
Acknowledgment	118
Author disclosure	118
References	118

Abstract

Antibody-dependent cellular phagocytosis (ADCP) is a process through which myeloid cells are able to exert their phagocytic function after recognition of opsonized bacteria, viruses, infected cells or any cells targeted by a specific antibody. ADCP of tumor cells represents a potent effector mechanism of monoclonal antibody therapy mediated by tumor associated macrophages (TAM) and other phagocytic cells as an *in situ* anti-tumor activity. Here we described a protocol based on flow cytometry and immunofluorescence assays enabling extensive comparative studies to address whether a monoclonal antibody engaging Fcγ receptors on macrophages can mediate *in vitro* ADCP of tumor cells.

List of abbreviations

ADCC	antibody-dependent cellular cytotoxicity
ADCP	antibody-dependent cellular phagocytosis
ANOVA	analysis of variance
APC	antigen presenting cells
BMDM	bone marrow-derived macrophages
CDC	complement dependent cytotoxicity
eTAA	extracellular tumor associated antigens
Fc	fragment crystallizable
FcR	fragment crystallizable receptor
Ig	immunoglobulin
IVT-RNA	in vitro transcribed ribonucleic acid
SEM	standard error of the mean
TAM	tumor associated macrophages

1 Introduction

Together with surgery, radiotherapy and chemotherapy, monoclonal antibody therapy is considered one of the main treatment of cancer therapy. Antibodies are able on one side to induce direct killing of tumor cells and on the other side to generate a long lasting immune response against the tumor-antigen expressing cells (Zahavi & Weiner, 2020). Monoclonal antibodies therapy treats cancer by mobilizing innate and adaptive immunity. Antibodies are large glycoproteins that belong to the immunoglobulin (Ig) superfamily and their primary role within the immune system is to recognize foreign antigens. They are composed of two heavy chains and two light chains, which pairs stabilized by disulfide bonds in a characteristic symmetric Y shape and composed by variable and constant regions. While the two N-terminal arms of the paired heavy and light variable regions are involved in the antigen recognition, as antigen binding portions, the C-termini of the 2 heavy chains (the base of the Y) constitute the fragment crystallizable (Fc) region, the fundamental part to mediate the interaction with other cellular and molecular components of the immune system (Pincetic et al., 2014). Fc regions are recognized by Fc receptors (FcR) present on a large variety of immune cells. Based on the type of heavy chains there are five different classes of antibodies: IgA, IgD, IgE, IgG and IgM. In the monoclonal antibody therapy, the IgGs are the most used ones, which interact with their specific receptors, the FcγR present on natural killer (NK) cells, neutrophils, monocytes, dendritic cells, and eosinophils. These receptors are able to mediate specialized functions such as antibody-dependent cellular cytotoxicity (ADCC) and complement-dependent cytotoxicity (CDC). Four different receptors can mediate ADCP, namely FcγRI (CD64), FcγRII (CD16), FcγRIII (CD32) and FcγRIV (CD16-2). The FcγRIII has an inhibitory function in mice, while the others have an activator function (Nimmerjahn & Ravetch, 2008). Each receptor has a specific affinity for the different IgG isotypes, modulating the dynamic of the ADCC/ADCP process. Phagocytosis is a general mechanism by which a cell engulf a particle of large dimension ($>5\,\mu m$) into the cell body (Rosales & Uribe-Querol, 2017). The binding of the Fc fragment to its receptor triggers an intracellular signaling cascade that leads to the engulfment (and killing) of the IgG opsonized particles or cells in a cytoskeleton dependent manner (Gordon, 2016; Russell, 2011). This process gives rise to an internal compartment known as phagosome, which fuses intracellularly and selectively with primary lysosomes or the product of the endoplasmic reticulum (ER) and Golgi complex, to form a secondary phagolysosome. ADCP reflects one mode of action of monoclonal therapeutic antibodies via the recognition of an extracellular tumor associated antigen (eTAA) overexpressed by cancer cells. This active process represents one of the most common activity of myeloid to attack opsonized tumor cells, as NK cells do via ADCC (Feng et al., 2019). In particular, antigen presenting cells (APC) after phagocytosis of tumor cells process and present the TAA derived epitopes to T cells initiating the adaptive immune response.

Human epidermal growth factor 2 (HER2) is a tyrosine kinase receptor that is overexpressed in many cancer types, particularly in ovarian and breast carcinomas (Slamon et al., 1989). After dimerization, HER2 receptors activate proliferative and anti-apoptotic intracellular signaling cascades (Chen, Lan, & Hung, 2003). Antibodies targeting HER2 achieve signaling perturbation by inhibiting dimerization and signaling activation. Trastuzumab was the first FDA approved anti-HER2 mAb and remains an important therapeutic option for primary and metastatic HER2-positive breast cancer patients (Albanell & Baselga, 1999). Trastuzumab is a humanized IgG1 monoclonal antibody, with a high affinity for FcγR and therefore a strong capacity to mediate ADCC in patients (Petricevic et al., 2013). Thus, the antigen recognition by the Fab region, which results in HER2 signaling inhibition, as well as the induction of immune effector functions such as ADCC mediated by the Fc region play a fundamental dual role in the mechanisms of action of trastuzumab (Clynes, Towers, Presta, & Ravetch, 2000; Musolino et al., 2008; Spector & Blackwell, 2009). Studies of ADCP conducted so far are mainly based on non-adherent lymphoid cells as target

(Kurdi et al., 2018; Shi et al., 2015), while here we describe a method to address whether a monoclonal antibody induces ADCP of adherent or suspension tumor cells. This workflow represents a universal tool to test phagocytosis of cancer cells *in vitro* by macrophages induced by any antibody-target receptor combination.

2 Materials

2.1 Disposable materials

1. 96-well round-bottom plate (Cornig 3799)
2. 50 mL and 15 mL Falcon (Greiner 227261, Greiner 188271)
3. Petridish 100 x 20 mm (Greiner cat: 633180) no treated dishes
4. Six-well plates (Falcon 353046)
5. Cell strainer (Corning 352350)
6. Pipet tips (Eppendorf)
7. 1.5 mL reaction tubes (Eppendorf 0030.121.589)
8. Serological pipet, 10 mL (Greiner Bio-One GmbH 607180)
9. Serological pipet, 5 mL (Greiner Bio-One GmbH 606180)
10. Serological pipet, 25 mL (Greiner Bio-One GmbH 760180)
11. T75 flasks (with filter) (Greiner Bio-One GmbH 658175)
12. Electroporation cuvettes (0.4 cm electrode VWR 732-1137)
13. Sterile round coverslips (Marienfeld 0117520)
14. 24-well round-bottom plates (Corning 3524)
15. 5 mL round-bottom tubes (Falcon 352350)
16. 5 mL tubes (Eppendorf 30119401)

2.2 Reagents

1. IMDM (Gibco 21980032)
2. RPMI 1640 (Life Technologies GmbH 61870036)
3. Glutamin (Gibco 25030-081)
4. ß-Mercaptoethanol (Gibco 31350-010)
5. Penicillin/Streptomycin (Gibco 15140-122)
6. Sodium Pyruvate (100 mM, Gibco 11360-039)
7. FBS (Sigma, F7524)
8. Blasticidin, 10 mg/mL (Invivogen ant-bl, BLL-38-05A)
9. EDTA 0.5 M (Sigma 03690)
10. Accutase (Life Technologies A1110501)
11. RNaseZap (Invitrogen AM9782)
12. X-VIVO 15, serum-free, hematopoietic cell medium Media (Lonza BE02-060Q)
13. DPBS, no calcium, no magnesium (Life Technologies 14190094)
14. M-CSF (Peprotech 315-02)

15. In vitro transcribed (IVT)-RNA encoding eGFP (Holtkamp et al., 2006)
16. Anti-human HER2, Trastuzumab, human IgG1 (InVivoGen BE0277 clone 7.16.4) for the opsonization of HER2$^+$ tumor cells
17. Isotype control (mouse IgG2a C1.18.4, In vivo Mab cat no: BE0085)
18. Fixable viability dye (L/D) eFluor 780 (eBioscience 65-0865-14 dilution 1:800)
19. Rat anti-mouse F4-80 BV421 (Biolegend 123132 dilution 1:100) for flow cytometry staining
20. Mouse anti-human/rat Her2 monoclonal antibody (Bio X Cell BE0277) for quality check by flow cytometry
21. F(ab')2-Goat anti-Mouse IgG (H+L)-A647 (Invitrogen, A-21237 dilution 1:500) for flow cytometry staining
22. Purified Rat anti-mouse (CD16/32 Mouse FC-Block BD 553142) for flow cytometry staining
23. BD Stabilizing reagent (BD, 339860) for flow cytometry staining
24. BD CompBeads Anti-Rat and Anti-hamster Ig, k (BD 51-90-9000949)
25. BD CompBeads Negative control (FBS) (51-90-900129)
26. Poly-L-lysine (Sigma, P8920)
27. PFA 32% (EMS, 1571-S) diluted to 4% in PBS for cell fixation
28. Triton X-100 (Sigma, T8787)
29. BSA (Applichem, A1391,0100)
30. Hamster anti-mouse CD11b-A647 (Biolegend, 101218 dilution 1:100) for IF staining
31. Phalloidin (ThermoFisher, A22283 dilution 1:200) for IF staining
32. Hoechst 33342 (Invitrogen, H3570 dilution 1:5000) for IF staining
33. Mounting media FlouromontG (eBioscience, 00-4958-02)

2.3 Equipments

1. Electroporator Gene Pulser II (BioRad) BTX, ECM830
2. Incubators Heracell 37 °C, 5% CO_2
3. Centrifuge 5810R (Eppendorf)
4. LSR Fortessa (BD Bioscience)
5. Confocal microscope Sp8 (Leica)

2.4 Cells and mice

1. MC38 (Hos et al., 2019) colon adenocarcinoma cells, kindly provided by Prof. F.A. Ossendorp, PhD, Leiden University Medical Center, Leiden, The Netherlands, were transduced to overexpress HER2 (data not published).
2. Femurs and tibias from C57BL6/J mice were used to isolated bone marrow hematopoietic cells as previously described (Kreiter et al., 2016).

2.5 Cell culture
1. MC38-Her2$^+$ cells were cultivated in tumor media (IMDM+2 mM L-glutamin, 50 μM ß-mercaptoethanol+8% v/v not heat inactivated FBS supplemented by blasticidin 1 mg/mL)
2. Bone marrow hematopoietic cells were cultivated in bone marrow differentiated macrophages (BMDM) medium (RPMI 1640 media+2 mM L-glutamin, 50 μM ß-Mercaptoethanol+8% v/v heat inactivated FBS and supplemented with 100 ng/mL M-CSF) for 7 days with a cell density of 4×10^5 cells/mL in Petri dishes (in 10 mL) to complete differentiation in macrophages. The cultures were fed on day 4 with new complete BMDM medium.

2.6 Software
1. BD FACS DIVA for flow cytometry acquisition
2. FlowJo v10.8.0 for flow cytometry analysis
3. GraphPad Prism 9.2.0 for graphical representation and statistical analysis
4. Leica SP8 for image acquisition
5. Fiji (ImageJ 1.53j) for image analysis

2.7 Methods
2.7.1 Cell culture
(i) The murine colon adenocarcinoma cells MC38, overexpressing Her2 (data not published) were cultivated in tumor medium (IMDM+2 mM L-glutamin, 50 μM ß-mercaptoethanol+8% v/v not heat inactivated FBS) in standard culture conditions (at 37 °C and 5% CO_2) in conventional 75-cm^2 cell culture flasks. Before full confluence was reached, cells were gently washed with 10 mL of pre-warmed PBS, detached with 2 mL accutase. The reaction was stopped by adding 8 mL of media to the cells. Cells were collected and centrifuged 300 x *g* for 8 min and counted in a Neubauer chamber.
(ii) As effector cells we used *in vitro* differentiated bone marrow derived macrophages (BMDM) generated (*Note 1*) in M-CSF (100 ng/mL) in BMDM in medium in standard culture conditions (37 °C, 5% CO_2) in 10 cm Petri dishes for 7 days. BMDM were harvested followed by an incubation for 10 min in cold 10 mM EDTA- PBS, and by a centrifugation at 300 x *g* for 8 min. Cells were counted in a Neubauer chamber.

2.7.2 MC38-Her2+ electroporation
Target cells were electroporated with eGFP-RNA (Diken, Kreiter, Selmi, Türeci, & Sahin, 2013) in order to be distinguished from the effector BMDM (*Note 2*) as following:

(i) Clean all pipettes, pipette tips as well as working space with RNaseZAP to eliminate RNases
(ii) Target MC38-Her2$^+$ cells were cultivated as described in Section 1.1. and in absence of blasticidin after electroporation
(iii) 5×10^6 cells were washed with x-Vivo medium (Lonza BE04-418Q) serum-free medium) and transferred to a 4 mm cuvettes in 250 μL X-Vivo medium

(iv) 2.5 µg IVT-RNA eGFP (Holtkamp et al., 2006) was added to the 4 mm cuvettes and mixed by gentry pipet up and down with a 200 µL pipette
(v) The cuvette was electroporated by applying 300 V with 1 pulse for 15 ms at room temperature
(vi) Cells were transferred in a six-well plate with the addition of 5 mL of tumor media (5×10^6/well) and incubated overnight in the incubator at 37 °C and 5% CO_2
(vii) After 24 h the MC38-Her2$^+$ cells were washed with warm PBS, treated with accutase to detach the cells and counted
(viii) Quality check by flow cytometry was perform to verify the eGFP expression (the reporter gene encoded by IVT-RNA) was performed 24 h after electroporation using 1×10^6 cells, stained with live dead reagent to quantify the viability of the cells (Fig. 1)
(ix) The rest of the cells were further used to perform the ADCP assay

2.7.3 ADCP assay
The ADCP assay mainly consist of two steps:

– The opsonization of the target cells using the specific antibody
– The co-culture of the target with the effector cells. During co-culture the effector cells recognize the opsonized cells as a target and engulf them exerting their effector function

The ADCP assay was performed by using flow cytometry and immunofluorescence analysis as read-outs. The steps of the two read-outs are slightly different from each other and therefore described in details in two different sections below.

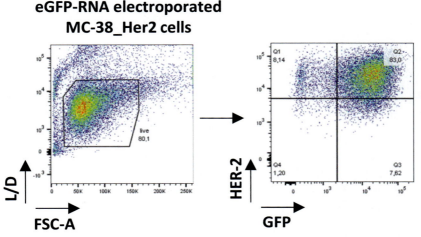

FIG. 1

Expression of eGFP (reporter gene) by flow cytometry performed 24 h after electroporation of MC38-Her2$^+$ cells with eGFP-RNA.

2.7.4 Flow cytometry
2.7.4.1 Opsonization of the target cells
i. 5×10^6 MC38-Her2$^+$ eGFP expressing cells were incubated with 5 μg/mL Trastuzumab (Her2neu) or the Isotype control in a 1.5 mL reaction tube in a final volume of 1.25 mL BMDM media *(Note 3)* at 37 °C for 10 min.

2.7.4.2 Phagocytosis assay
ii. BMDM were seeded in a 96-well round-bottom plate at a density of 0.5×10^5/well in 50 μL and let adhere for half an hour in the incubator at 37 °C and 5% CO_2.
iii. Target cells (in suspension) at density of 1×10^5/well in 50 μL were added to the adherent BMDM to reach a target: effector ratio of 2:1
iv. To allow recognition of the target and foster phagocytosis, cancer cells-BMDM co-culture were maintained for 2 h in the incubator at 37 °C and 5% CO_2 *(Note 4)*.
v. Cells were harvested and stained with life dead first for 10 min at 4C in FACS buffer and after washing with FACS buffer, centrifuged and stained with anti- mouse F4-80 antibody for 20 min at 4C. After washing and centrifugation, cells were further analyzed by flow cytometry (Fig. 2) *(Note 5)*.

2.7.5 Immunofluorescence and confocal imaging
2.7.5.1 Cover slip coating
i. Single glass coverslips were inserted in each well of a 24 well plate and coated with 500 μL Poly-L-Lysine, 0.1%, dilute in Millipore water 1:10 for 60 min at RT
ii. After the incubation time, Poly-L-Lysine solution was aspirated
iii. The coverslip were rinsed three times with 1 mL ddH_2O
iv. Coverslip were dry for 30 min at RT under sterile conditions

2.7.5.2 Co-culture
(i) 3×10^5 BMDM were seeded on a coated coverslip in a volume of 300 μL and incubated 1 h at 37 °C and 5% CO_2 to allow full adhesion on the coverslips
(ii) 6×10^5 opsonized target cells in volume of 300 μL (in suspension as described in Sections 3.1.1 and 3.1.2) were added to the effector cells in a ratio target: effector 2:1 and co-cultured for 4 h at 37 °C 5% CO_2 in a final volume of 600 μL
(iii) Supernatants were aspirated and cells fixed for 10 min in PFA 4% followed by the staining with anti-mouse CD11b-A647 and Phalloidin-A546 for 1 h in PBS containing 0.1% Triton and 0.5% BSA. Coverslips were washed extensively with 0.5% BSA-PBS and nuclei staining performed using Hoechst diluted in PBS. After washing with PBS coverslips were mounted on slides (Fig. 3).
(iv) Images were acquired at the confocal microscope and processed with Image J 1.53j software.

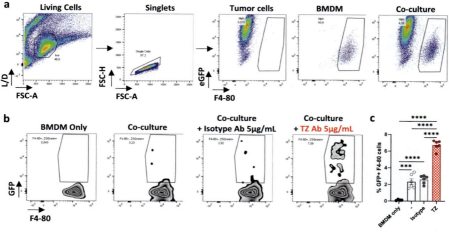

FIG. 2

ADCP assay (A) gating strategy of samples acquired by the flow cytometer. The samples, such as MC38-Her2$^+$ eGFP$^+$ target cells, BMDM and the co-cultured cells, are displayed. (B) Representative Zebra plots of the F4-80$^+$ gated BMDM. The indicated gate includes the F4-80$^+$ BMDM that have engulfed the target MC38-Her2$^+$ eGFP$^+$ cells opsonized by Trastuzumab (TZ) or treated with isotype antibody control. (C) Histogram shows the percentage of eGFP$^+$ F4-80$^+$ BMDM under the different conditions. Statistical analysis performed using one-way ANOVA and Tukey's multiple comparison test with *$P \leq 0.05$, **$P \leq 0.01$, ***$P \leq 0.001$, ****$P \leq 0.0001$.

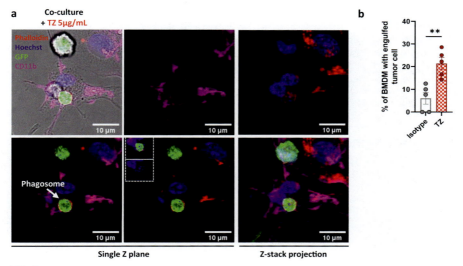

FIG. 3

ADCP Assay (A) representative immunofluorescence images of BMDM (magenta, CD11b$^+$) and MC38-Her2+ eGFP+ cell (green, eGFP) treated with TZ. BMDM has phagocyted an opsonized target cell after 4h of co-culture. The white arrow indicates the phagosome defined by a strong actin staining (red, phalloidin) and a different z-plane is shown in the white boxes, where is possible to observe the fragmented nuclei (blue, Hoechst) of the engulfed MC38-Her2$^+$ eGFP$^+$ cell. Images of a single Z-plan (left) and the Z-stack projection (right) are represented. (B) Percentage of engulfed tumor cells by macrophages is plotted. BMDM in 5 fields of view were analyzed (isotype $n=50$, TZ $n=33$). Data are shown as mean±SEM using unpaired t-test ($P^{**}=0.0032$).

3 Notes

Note 1: BMDM are generated prepared following the published protocol (Trouplin et al., 2013).

Note 2: Transiently or stably transfected fluorescent cells or fluorescently label cells (stained with fluorescent cell tracker dyes) can alternatively be used and analyzed by flow cytometry as well as by immunofluorescence. In our protocol, we favored the IVT-RNA electroporation to label the target cells instead of the fluorescent cell trackers, since we noticed that the fluorescent dyes can leak and stain the BMDM in case very extensive repetitive washing steps are not well performed. Therefore IVT-RNA electroporation reduced dramatically the unspecific labelling of BMDM.

Note 3: A titration of the specific antibody might be necessary in order to find the optimal working concentration.

Note 4: Different co-culture time (3, 4 or 6 h) can be tested in order to optimize the assay and the specific experimental settings.

Note 5: Gate strategy of the flow cytometry analysis of engulfed eGFP+ target cells in F4-80+ BMDM is defined by four sequential steps (Fig. 2). First, exclusion of the large fraction of live cells (FCS-A > 130 K, which include doublets) and inclusion of the cells indicated in the gate in the first dot plot (live/Dead vs FSC-A) (Fig. 2A). Next, inclusion of single cells with a second stringent gate as indicted in the second dot plot (FSC-H vs FSC-A, singlet gate) and exclusion of possible doublets (Fig. 2A). Third, BMDM are gated as F4-80+ cells (Fig. 2A). Last, phagocytic BMDM with engulfed eGFP+ target cells are gated as F4-80+eGFP+ cells (Figs. 2B and 3).

Acknowledgment

The authors thank Biopharmaceutical New Technologies (BioNTech) SE for proving eGFP encoding IVT-mRNA; Lena Haas for her technical assistance.

Author disclosure

Authors have no conflicts of interest to disclose.

References

Albanell, J., & Baselga, J. (1999). Trastuzumab, a humanized anti-HER2 monoclonal antibody, for the treatment of breast cancer. *Drugs of Today (Barcelona, Spain: 1998)*, 35(12), 931–946.

Chen, J.-S., Lan, K., & Hung, M.-C. (2003). Strategies to target HER2/neu overexpression for cancer therapy. *Drug Resistance Updates: Reviews and Commentaries in Antimicrobial and Anticancer Chemotherapy*, 6(3), 129–136. https://doi.org/10.1016/s1368-7646(03)00040-2.

Clynes, R. A., Towers, T. L., Presta, L. G., & Ravetch, J. V. (2000). Inhibitory Fc receptors modulate in vivo cytotoxicity against tumor targets. *Nature Medicine*, 6(4), 443–446. https://doi.org/10.1038/74704.

Diken, M., Kreiter, S., Selmi, A., Türeci, O., & Sahin, U. (2013). Antitumor vaccination with synthetic mRNA: Strategies for in vitro and in vivo preclinical studies. *Methods in Molecular Biology (Clifton, N.J.)*, 969, 235–246. https://doi.org/10.1007/978-1-62703-260-5_15.

Feng, M., Jiang, W., Kim, B. Y. S., Zhang, C. C., Fu, Y.-X., & Weissman, I. L. (2019). Phagocytosis checkpoints as new targets for cancer immunotherapy. *Nature Reviews. Cancer*, 19(10), 568–586. https://doi.org/10.1038/s41568-019-0183-z.

Gordon, S. (2016). Phagocytosis: An immunobiologic process. *Immunity*, 44(3), 463–475. https://doi.org/10.1016/j.immuni.2016.02.026.

Holtkamp, S., Kreiter, S., Selmi, A., Simon, P., Koslowski, M., Huber, C., et al. (2006). Modification of antigen-encoding RNA increases stability, translational efficacy, and T-cell stimulatory capacity of dendritic cells. *Blood*, 108(13), 4009–4017. https://doi.org/10.1182/blood-2006-04-015024.

Hos, B. J., Camps, M. G. M., van den Bulk, J., Tondini, E., van den Ende, T. C., Ruano, D., et al. (2019). Identification of a neo-epitope dominating endogenous CD8 T cell responses to MC-38 colorectal cancer. *Oncoimmunology*, 9(1), 1673125. https://doi.org/10.1080/2162402X.2019.1673125.

Kreiter, S., Diken, M., Selmi, A., Petschenka, J., Türeci, Ö., & Sahin, U. (2016). FLT3 ligand as a molecular adjuvant for naked RNA vaccines. *Methods in Molecular Biology (Clifton, N.J.)*, 1428, 163–175. https://doi.org/10.1007/978-1-4939-3625-0_11.

Kurdi, A. T., Glavey, S. V., Bezman, N. A., Jhatakia, A., Guerriero, J. L., Manier, S., et al. (2018). Antibody-dependent cellular phagocytosis by macrophages is a novel mechanism of action of elotuzumab. *Molecular Cancer Therapeutics*, 17(7), 1454–1463. https://doi.org/10.1158/1535-7163.MCT-17-0998.

Musolino, A., Naldi, N., Bortesi, B., Pezzuolo, D., Capelletti, M., Missale, G., et al. (2008). Immunoglobulin G fragment C receptor polymorphisms and clinical efficacy of trastuzumab-based therapy in patients with HER-2/neu-positive metastatic breast cancer. *Journal of Clinical Oncology: Official Journal of the American Society of Clinical Oncology*, 26(11), 1789–1796. https://doi.org/10.1200/JCO.2007.14.8957.

Nimmerjahn, F., & Ravetch, J. V. (2008). Fcgamma receptors as regulators of immune responses. *Nature Reviews. Immunology*, 8(1), 34–47. https://doi.org/10.1038/nri2206.

Petricevic, B., Laengle, J., Singer, J., Sachet, M., Fazekas, J., Steger, G., et al. (2013). Trastuzumab mediates antibody-dependent cell-mediated cytotoxicity and phagocytosis to the same extent in both adjuvant and metastatic HER2/neu breast cancer patients. *Journal of Translational Medicine*, 11, 307.

Pincetic, A., Bournazos, S., DiLillo, D. J., Maamary, J., Wang, T. T., Dahan, R., et al. (2014). Type I and type II Fc receptors regulate innate and adaptive immunity. *Nature Immunology*, 15(8), 707–716. https://doi.org/10.1038/ni.2939.

Rosales, C., & Uribe-Querol, E. (2017). Phagocytosis: A fundamental process in immunity. *BioMed Research International*, 2017, 9042851. https://doi.org/10.1155/2017/9042851.

Russell, D. G. (2011). Mycobacterium tuberculosis and the intimate discourse of a chronic infection. *Immunological Reviews*, *240*(1), 252–268. https://doi.org/10.1111/j.1600-065X.2010.00984.x.

Shi, Y., Fan, X., Deng, H., Brezski, R. J., Rycyzyn, M., Jordan, R. E., et al. (2015). Trastuzumab triggers phagocytic killing of high HER2 cancer cells in vitro and in vivo by interaction with Fcγ receptors on macrophages. *Journal of Immunology (Baltimore, Md.: 1950)*, *194*(9), 4379–4386. https://doi.org/10.4049/jimmunol.1402891.

Slamon, D. J., Godolphin, W., Jones, L. A., Holt, J. A., Wong, S. G., Keith, D. E., et al. (1989). Studies of the HER-2/neu proto-oncogene in human breast and ovarian cancer. *Science (New York, N.Y.)*, *244*(4905), 707–712. https://doi.org/10.1126/science.2470152.

Spector, N. L., & Blackwell, K. L. (2009). Understanding the mechanisms behind trastuzumab therapy for human epidermal growth factor receptor 2-positive breast cancer. *Journal of Clinical Oncology: Official Journal of the American Society of Clinical Oncology*, *27*(34), 5838–5847. https://doi.org/10.1200/JCO.2009.22.1507.

Trouplin, V., Boucherit, N., Gorvel, L., Conti, F., Mottola, G., & Ghigo, E. (2013). Bone marrow-derived macrophage production. *Journal of Visualized Experiments: JoVE*, *81*, e50966. https://doi.org/10.3791/50966.

Zahavi, D., & Weiner, L. (2020). Monoclonal antibodies in cancer therapy. *Antibodies (Basel, Switzerland)*, *9*(3). https://doi.org/10.3390/antib9030034.

CHAPTER

Quantification of lymphocytic choriomeningitis virus specific T cells and LCMV viral titers

9

Melanie Grusdat[a,b], Catherine Dostert[a,b], and Dirk Brenner[a,b,c],*

[a]*Experimental and Molecular Immunology, Department of Infection and Immunity, Luxembourg Institute of Health, Esch-sur-Alzette, Luxembourg*
[b]*Immunology & Genetics, Luxembourg Centre for Systems Biomedicine, University of Luxembourg, Esch-sur-Alzette, Luxembourg*
[c]*Odense Research Center for Anaphylaxis (ORCA), Department of Dermatology and Allergy Center, Odense University Hospital, University of Southern Denmark, Odense, Denmark*
*Corresponding author: e-mail address: Dirk.Brenner@lih.lu

Chapter outline

1 Introduction	122
2 Materials	123
3 Methods	124
3.1 Tetramer production	124
3.2 Tetramer staining	124
3.3 Cytokine production of virus-specific T cells	125
3.4 Plaque assay (plaque forming unit assay)	126
3.4.1 Day 1 seeding MC-57 cells	127
3.4.2 Day 3 staining of the plaque assay	128
4 Notes	129
5 Conclusion	129
Acknowledgments	130
References	130

Abstract

Lymphocytic choriomeningitis virus (LCMV) is a frequently used animal model to study immune responses against acute and chronic viral infections. LCMV is a non-cytopathic virus, but destruction of infected cells by cytotoxic T lymphocytes (CTLs) can lead to severe damage of tissues. Virus-specific T cell responses have to be balanced. A low virus load leads to a strong T cell response and subsequently to viral control. In contrast, a high viral titer is associated with T cell exhaustion and chronic viral infections. During an intermediate LCMV viral load $CD8^+$ T cells can cause immunopathology, which can have detrimental outcomes. The LCMV infection model offers the opportunity to study virus-specific $CD4^+$ and $CD8^+$ T cell responses during chronic and acute infections by quantifying LCMV-specific T cells by tetramer staining and measuring cytokine production and viral titers in different organs.

1 Introduction

Antigen-specific T cells are an important arm of the adaptive immune system. They are the essential players in cell-mediated immunity. Lymphocytic choriomeningitis virus (LCMV) is widely used to study immunity and especially to examine immune responses against acute and chronic infections (Althaus, Ganusov, & De Boer, 2007). LCMV belongs to the arenavirus family and is a non-cytopathic virus, which causes immunopathology mainly by virus-specific $CD8^+$ T cells that target infected cells (Fung-Leung, Kundig, Zinkernagel, & Mak, 1991; Zinkernagel et al., 1986). The LCMV models has led to fundamental discoveries including major histocompatibility complex (MHC) restriction as well as T cell exhaustion (Wherry, Blattman, Murali-Krishna, van der Most, & Ahmed, 2003). The findings by Peter Doherty and Rolf Zinkernagel regarding MHC restriction of T cells were awarded the Nobel prize in Physiology or Medicine in 1996 (Doherty & Zinkernagel, 1975; Zinkernagel & Doherty, 1974, 1975). Oldstone and Dixon highlighted that a specific antibody response depends on mouse and virus strains (Hangartner, Zinkernagel, & Hengartner, 2006; Zhou, Ramachandran, Mann, & Popkin, 2012). In 1994, Lau et al. could show that memory cells are maintained in the absence of a virus. It was demonstrated that an adoptive T cell transfer of virus-specific T cells into uninfected mice could protect these naïve animals against a LCMV challenge (Lau, Jamieson, Somasundaram, & Ahmed, 1994). Ohashi et al. established a model of viral-induced autoimmunity (RIP-gp mouse), where cytotoxic T lymphocytes (CTLs) fail to discriminate between self and non-self (Ohashi et al., 1991; von Herrath, Guerder, Lewicki, Flavell, & Oldstone, 1995). In 2005, Barber et al. demonstrated that immunotherapy with programmed cell death protein 1 (PD1) blocking antibodies led to restoration of exhausted T cells during chronic infections (Barber et al., 2006). Nowadays, the LCMV infection model is still an important tool to study virus-specific T cell responses partially due to available excellent methods and tools to follow LCMV-specific T cell responses. Several different MHC class I and II-restricted epitopes are identified for LCMV, whereby for CTLs glycoprotein gp33 (gp33-H-2Db) and np396 (np396-H-2Db) are the immunodominant epitopes and for CD4$^+$ T cells it is gp66 (Gp66-I-A(b)). How to produce viral stocks of LCMV and the associated biosafety considerations are well described by Welsh et al. (Welsh & Seedhom, 2008). Here, we provide the fundamental methods to study LCMV-specific T cell responses, including the identification of LCMV-specific T cells by tetramer staining, re-stimulation with LCMV specific peptides and measuring LCMV titers in organs (Battegay et al., 1991; Gallimore et al., 1998; Murali-Krishna et al., 1998). We describe tetramer staining for flow cytometry, which can also be used to stain for virus-specific T cells in histological sections to highlight the immunological T cell synapse (Fooksman et al., 2010; McGavern, Christen, & Oldstone, 2002).

2 Materials

1. Antibodies for flow cytometry: anti-CD16/32 (clone: 93), anti-CD8α (clone: 53-6.7), anti-CD4 (clone: GK1.5), anti-CD279 (PD-1) (clone: 29F.1A12), anti-CD127 (clone: JES6-5H4), anti-CD44 (clone: IM7), anti-KLRG1 (clone: 2F1/KLRG1), anti-CD62L (clone: MEL-14), anti-CD152 (CTLA4) (clone: UC10-4B9), anti-CD336 (Tim-3) (clone: RMT3-23), anti-IFN-γ (clone: XMG1.2), anti-TNF (clone: MP6-XT22), IL-2 (clone: JES6-5H4), anti-CD107a (clone: eBio1D4B (1D4B))
2. Solutions for flow cytometry: BD Cytofix/Cytoperm™ (BD Biosciences, 554714), BD GolgiPlug™ BD Biosciences, 555029), Streptavidin APC (Biolegend, 405207), Zombie NIR™ Fixable Viability Kit (Biolegend, 423106), FACS buffer: 1x Phosphate buffered saline (PBS) supplemented 1% FCS and 1% 0.5M EDTA (EDTA adjusted to pH 8; Ethylenediaminetetraacetic acid (Sigma-Aldrich, I7633-10L))
3. biotinylated Monomers: H-2D(b) gp33 (KAVYNFATM), H-2D(b) np396 (FQPQNGQFI), I-A(b) gp66 (DIYKGVYQFKSV) stored at −80 °C, kindly provided by NIH Tetramer Facility
4. LCMV-peptides: GP$_{33-41}$ KAVYNFATM, NP$_{396-404}$ FQPQNGQFI, GP$_{66-77}$ DIYKGVYQFKSV (GenScript (Hong Kong))
5. Culture media and reagents: Media for MC-57 cells: Minimum Essential Medium Eagle, alpha mod (α-MEM media), (Sigma-Aldrich, M8042-6X500ML) supplemented with 1% P/S (Gibco, 11548876) and 5% FCS (Biochrom GmbH, S0615) for growing MC-57 cells, supplemented with 1% P/S and 2% FCS for Plaque assay; for 10XIMDM: Iscove's Modified Dulbecco's Medium, (Sigma-Aldrich, EDS-500G); Methyl Cellulose: Methocel® A15 LV (Sigma-Aldrich, 64605-500G-F); single cell solutions are cultured in IMDM with HEPES and L-Glutamine, (Westburg, LO BE12-722F) supplemented with 1% P/S and 10% FCS; overlay Media: Mix 1:1 (for each assay, mix only what is needed) 2XIMDM with 10%FCS, 2X P/S (Adjust pH) and 2%Methyl Cellulose

(10 g in 500 mL ddH$_2$O in a 1 L bottle, Stir until dissolved (o/n) at 4 °C, autoclave to sterilize and return to stir at 4 °C until dissolved); blocking solution: 10% FCS in 1xPBS
6. Antibodies for Plaque assay: Primary antibody: VL-4 supernatant produced by ourselves (see Notes) diluted in PBS with 1%FCS, secondary antibody: Peroxidase AffiniPure Goat Anti-Rat IgG (H+L) (JACKSON IMMUNORESEARCH EUROPE LTD, 112-035-003) (resuspend powder in 2 mL dH$_2$O and prepare aliquots of 50 µL. Store at −20 °C, 1 aliquot/10 plates in PBS with 1%FCS)
7. Reagents for Plaqueassay: OPD Substrate for the color reaction, Stock A: 0.2 M Na$_2$HPO$_4$*2H$_2$O (Sigma-Aldrich, 71645-1 kg), 35.6 g/L, Stock B: 0.1 M Citric acid (Sigma-Aldrich, C0759-500G), 19.2 g/L, OPD: Ortho-Phenylendiamin (Sigma-Aldrich, P8412), 30 mg/tablet, 30% H$_2$O$_2$ Perhydrol (Sigma-Aldrich, H1009-100 mL) Working buffer (50 mL, for 10 plates): Stock A: 12.5 mL, Stock B: 12.5 mL, ddH$_2$O: 25 mL, OPD: 30 mg, 30% H$_2$O$_2$: 50 µL (add only when tablets dissolved)
8. Equipment: BD-LSRFortessa™ (BD Biosciences), Qiagen TissueLyser II (Qiagen, 85300)
9. Disposable material: 24 well plates (Greiner bio-one, 662160), 96 well plates (greiner bio-one, 650185), flasks T-75 (greiner bio-one, 658175), flasks T-175 (greiner bio-one, 660175), Stainless Steel Beads, 5 mm (Qiagen, 69989)

3 Methods

3.1 Tetramer production

1. Let the monomers adapt to RT and take an appropriate aliquot (9 µL for np396 (2 mg/mL) or gp66 (1.7 mg/mL) and 4.5 µL for gp33 (2 mg/mL)).
2. Prepare 1:10 dilution of Streptavidin in FACS buffer (e.g. 47 µL Streptavidin and 423 µL FACS buffer/150 µL needed per tetramer).
3. Add 15 µL of Step2 to the monomers and mix carefully by pipetting up and down. Let it incubate for 5 min. Repeat this step four times (in total add 60 µL).
4. Add 30 µL of Step 2 to the monomers and mix carefully by pipetting up and down. Let it incubate for 5 min. Repeat this step three times (in total add 90 µL).
5. Store at 4 °C for up to 4 months (protected from light).

3.2 Tetramer staining

1. Preparation of samples: Collect spleen in 5 mL of IMDM media supplemented with 10% FCS and 1% antibiotics.
2. Smash spleen in a 70 µm cell strainer, count the cells and adjust cell count to 1×10^7 per mL.

3. Aliquot 1×10^6 cells in a 96 well plate per staining and centrifuge plate at $350 \times g$ for 5 min at 4 °C.
4. Optional step: To avoid unspecific antibody binding an Fc-blocking step can be performed prior the tetramer-staining. Incubate the cells with a purified anti-CD16/32 antibody (clone: 93) at the recommended dilution in 25 µL for 15 min at 4 °C (add the tetramer in 25 µL FACS buffer instead of 50 µL).
5. Resuspend cells in 50 µL of freshly prepared tetramer-staining (1 µL per staining in FACS buffer; see Notes 4.1) per well and incubate for 15 min for gp33 and np396 or 30 min for gp66 at 37 °C in the incubator.
6. Add additionally surface staining in FACS buffer and incubate for further 30 min at 4 °C (e.g. CD8, CD4, Zombie NIR™ Fixable Viability, PD1, CD127, CD44, KLRG1, CD62L, CTLA4, TIM3).
7. Wash cells with 200 µL FACS Buffer and centrifuge plate at $350 \times g$ for 5 min at 4 °C.
8. Resuspend in FACS Buffer and analyze samples by flow cytometry. Note: We recommend to include a fixation step (e.g. 15 min at RT with 2–4% Formaldehyde; followed by two washes with FACS Buffer). Follow the local health and safety laboratory measures regarding handling LCMV (see Fig. 1A for exemplary gating strategy).

3.3 Cytokine production of virus-specific T cells

1. Aliquot 1×10^6 cells in a 96 well plate (from cells prepared in step 2 from Section 3.2) per staining and prepare for each peptide an additional well for media/DMSO control for CD4 and CD8 staining (five wells in total) and centrifuge plate at $350 \times g$ for 5 min at 4 °C.
2. Resuspend cells in 100 µL of IMDM media (see Materials 2.5) and add additionally 100 µL of LCMV peptides gp33, np396 (diluted to 10 µg/mL in IMDM media; final concentration: 5 µg/mL) to the cells or media/DMSO control and incubate for 1 h at 37 °C in the incubator.
3. Add 20 µL of 1:100 pre-dilution BD GolgiPlug™ in IMDM media to each well (final dilution 1:1000) and incubate for 5 h at 37 °C in the incubator. After the incubation centrifuge plate at $350 \times g$ for 5 min at 4 °C.
4. Resuspend cells in 50 µL per well of surface staining in FACS buffer and incubate for 30 min at 4 °C (or leave in this step overnight) (e.g. CD8, CD4, Zombie NIR™ Fixable Viability).
5. Wash cells with 200 µL FACS Buffer and centrifuge plate at $350 \times g$ for 5 min at 4 °C.
6. Fix cells in 100 µL BD Cytofix/Perm for 20 min at 4 °C and centrifuge plate at $350 \times g$ for 5 min at 4 °C.
7. Wash 2x in 1x BD Permwash.
8. Add intracellular staining (in Permwash) and incubate for 30 min at 4 °C (e.g. IFNγ, Il-2, TNF).

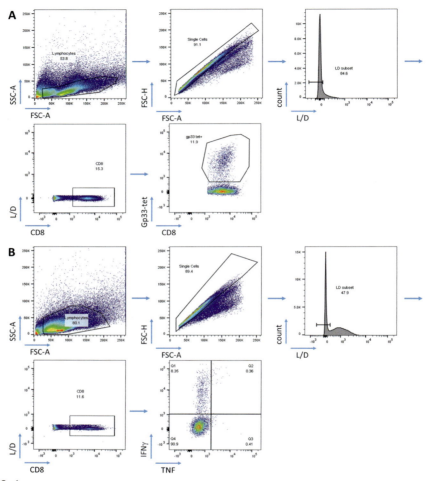

FIG. 1

Exemplary flowcytometry gating strategy shown for a Day 12 p.i. of LCMV-WE 2×10^6 infected C57/Bl6 mice. (A) gp-33 tetramer; (B) intracellular IFNγ and TNF after gp33 peptide restimulation.

9. Wash cells with 200 μL FACS Buffer and centrifuge plate at $350 \times g$ for 5 min at 4 °C.
10. Resuspend in FACS Buffer and analyze samples by flow cytometry (see Fig. 1B for exemplary gating strategy).

3.4 Plaque assay (plaque forming unit assay)

Plaque forming unit assays are a golden standard for quantification of virus titers e.g. LCMV through counting of plaques (infectious units) macroscopically. On day 1 MC-57 cells are infected by serial dilution with the samples containing unkown

LCMV virus concentration. The monolayer of cells is covered with an overlay to prevent viral spread through the media and allowing the formation of individual plaques by only infecting surrounding cells. On day 3 the formed plaques are visualized by immunostaining with a monoclonal rat anti-LCMV antibody and a peroxidase-labeled second antibody step allowing visual counting of brown plaques by eye.

3.4.1 Day 1 seeding MC-57 cells

1. Prepare round bottom 96 well plates for dilution of the samples by adding 130 μL of MEM 2%FCS to rows 2–12 if plaquing organ samples or cell culture supernatants and to rows B–G in case of plaquing blood samples, label 96 well plates with increasing numbers. Label the same amount of 24 well plates with increasing numbers.
2. Calculate the number of MC-57 cells needed: 5 mL of cells are needed for one plate, which corresponds to four samples of organs or supernatants or six samples of blood. Prepare the MC-57cells at a concentration of 8×10^5/mL and keep on ice. Invert the tube occasionally to keep the cells in suspension. Tip: It is preferable to aliquot them in sterile 50 mL tubes rather than in Duran® laboratory reusable bottles because the cells may attach to the side or bottom of the flask.
3. Defreeze supernatant (SN) or organs. For organ homogenization, organ pieces have to be collected in 2 mL Eppendorf Safe-Lock microcentrifuge tubes with 1 mL of α-MEM media and a steel bead. Lyse organs with Qiagen TissueLyser II, "frequency 1/s" = 03.0 and "Time min/sec" = 3.00. Keep organs on ice.
4. Add 90 μL of sample or supernatant to the first row in duplicates. For serum, dilute 1:4 or 1:5 in MEM for titration and add in duplicates to row A. Tip: Liver samples can be hard to pipette. In case of problematic pipetting samples can be either shortly centrifuged down and supernatants are only used or air of pipette is released at the bottom of the tube and liquid is quickly taken up.
5. Make the serial dilutions with a multi-channel pipette by taking 60 μL from the first row, adding it to the second row, mix samples four to six times and carrying over 60 μL over to the next row. Change tips in the middle of the plate.
6. Add 200 μL of the samples to the well in the corresponding 24 well plate, working with a multi-channel pipette. This is easiest accomplished by setting the pipette to 100 μL and placing two tips in each well e.g. pool row A and B (samples in duplicates) of the 96 well plate in well A of the 24 well plate. Add samples backwards from rows 12 to 6, 10 to 5, 8 to 4, 6 to 3, 4 to 2, 2 to 1 or from rows H, F, D and B for serum. It is not necessary to change tips for this procedure.
7. Using a multi-channel pipette, add 200 μL of MC-57 cells to each well of the 24 well plates. Leave the pipette at 100 μL and use the 12 channels, to avoid touching the plate at the same spot as the samples. Use a reservoir for the cells. Tip: Pipette with the positive pipette method: press multi-channel down to second pressure point (more than 100 μL in each tip) and release until first pressure point. Thereby you reduce risk of uneven pipetting while reusing tips.

8. Tap the plate to mix the cells and the sample and put it in a CO_2 incubator for 4 h or until the cells have settled on the bottom of the well. The cells can be round, but they should not be floating.
9. Add 200 μL of the overlay (see Materials 2.5) with a multichannel using the same positive pipette method technique.
10. Return the plates to the incubator and leave them for 2 days. Check the cells after 1 day to make sure the cells are growing.

3.4.2 Day 3 staining of the plaque assay

1. On day 3 plates should be checked under the microscope to make sure there is a confluent MC-57 monolayer.
2. Gently flick off the overlay into a virus waste container. Wear eye and mouth protection to protect from virus (follow the local health and safety laboratory measures regarding handling LCMV). Take care not to let the monolayer dry out before it is fixed. Collect SN in appropriate biosafety garbage.
3. Under the hood, add 200 μL of 4% Formaldehyde-PBS to each well to fix cells. Incubate for 30 min. The next steps can be performed outside of the biosafety cabinet after the cells are fixed and plates are sterilized from outside. All chemicals should be disposed of in appropriate garbage.
4. Wash 2x with PBS. (All washing steps can be flicked off down the drain.). Tip: For the washing steps plates can be submerged in a bucket containing 1xPBS.
5. Add 200 μL 1% Triton-X100 in PBS. Incubate for 20 min. Tip: Start the timer when you are done with the first plate. Take care of the order and start washing the first ones exactly after 20 min, in the same order and same speed as putting the Triton solution.
6. Wash 2x with PBS.
7. Add 200 μL PBS with 10% FCS to block non-specific binding. Incubate for 1 h at RT.
8. Add 200 μL VL-4 Rat anti-LCMV mAb diluted in PBS with 1% FCS (self-produced VL-4 antibody, dilution has to be tested for each batch). Incubate for 1 h at RT.
9. Wash 2x with PBS.
10. Add 200 μL per well of secondary antibody diluted in PBS with 1% FCS. Incubate for 1 h at RT.
11. Wash 2x with PBS.
12. Add 200 μL per well of the color reaction working buffer. Incubate for 15–20 min until a good color is developed and plaques are clearly identifiable (plaques brown, background not yet too yellow).
13. Dispose SN in appropriate garbage container and wash the plates under running water in a container to remove the excess dye solution. Dry inverted on absorbent paper.
14. Let the plates dry and count plaques ideally the day itself.

Calculation:
   Count number of plaques. Multiply this number by 3 (as more or less ⅓ of every organ is taken) and 10 for dilution factor on plate. Supernatants and blood sample calculation depends on how they are pre-diluted.

4 Notes

1. For the tetramer production use sterile FACS buffer and switch off light in the hood.
2. Aliquot monomers in appropriate volume to avoid freeze and thaw cycles.
3. Tetramer staining can be also performed on blood and liver samples.
4. Tetramer concentration needs to be evaluated with every fresh prepared batch (recommend to test 0.25–2 µL per stain).
5. LCMV peptides should be diluted to 1 mg/mL in DMSO.
6. Degranulation of LCMV-specific T cells can be determined by incubating the splenocytes with anti-CD107a antibodies (LAMP-1) during step 3.3. cytokine production of virus-specific T cells. Anti-CD107a is added with the LCMV-specific peptides or DMSO control in step 3.3.2. 1h prior addition of BD GolgiPlug™.
7. MC57G cell line (MC-57 cells) is a murine C57/Bl6-derived methylcholanthrene-induced fibrosarcoma cell line, which is a standard cell line for murine viral immunity studies. Take MC-57 cells at least 10 days in culture before the plaque assay. For 10 plates one T175 confluent flask is needed. MC-57 cells are adherent cells and have to be trypsinized. Culture media for MC-57 cells: MEM 5%FCS 1% penicillin/streptomycin.
8. Plaque assay: for the washing step use a bucket for PBS and submerge the plates.
9. Organs to use for plaque assay: The most commonly analyzed organs for titers are: spleen (one third), liver (one lobe only, usually the largest), kidney, lung (1 lobe), spinal cord and brain. For blood: sera or plasma can be analyzed.
10. VL-4 cell culture: Thaw VL-4 cells and culture them in a T-75 flask with 15 mL DMEM 10%FCS 1%P/S. On day 2, re-seat them in a T-175 flask with 35 mL fresh medium (pellet cells before). On day 4, split them into 3 T-175 flasks and add medium up to 150 mL per flask. Collect supernatant when medium turns orange-yellow (more or less 1 week). Centrifuge at 2000 rpm for 20 min, pool supernatant in a fresh flask. Aliquot SN and store at −20 °C. Test each SN in Plaque assay.

5 Conclusion

The present protocols describe detailed methods how to prepare tetramers for identification of LCMV-specific T cells and how to utilize these tetramers. Combining a tetramer staining with different T cell markers can visualize the activation status of the virus-specific T cells, including T cell exhaustion, memory T cells and short-lived effector T cells. Furthermore, we describe how to measure cytokine production of LCMV-specific T cells. This method can be also used to analyze degranulation markers e.g. LAMP-1 (anti-CD107a) staining. Furthermore, we describe how to determine virus titers. Overall, we conclude that LCMV is an excellent model to study in detail the development of T cell responses over the course of infections. Additionally, different strains of LCMV can be utilized to study both chronic (e.g. Clone-13, DOCILE) and acute infections (e.g. Armstrong, WE).

Acknowledgments

The authors would like to thank the National Cytometry Platform, the Luxembourg Institute of Health's Animal Welfare Structure and the National Institutes of Health (NIH) Tetramer Core Facility. The cell lines VL-4, MC-57 and the LCMV strains, that have been used to establish this protocol, were kindly provided by Philipp A. Lang (University of Düsseldorf, Germany). The authors are supported by the FNRS Televie program (7459719F).

References

Althaus, C. L., Ganusov, V. V., & De Boer, R. J. (2007). Dynamics of CD8+ T cell responses during acute and chronic lymphocytic choriomeningitis virus infection. *Journal of Immunology*, *179*, 2944–2951.

Barber, D. L., Wherry, E. J., Masopust, D., Zhu, B., Allison, J. P., Sharpe, A. H., et al. (2006). Restoring function in exhausted CD8 T cells during chronic viral infection. *Nature*, *439*, 682–687.

Battegay, M., Cooper, S., Althage, A., Banziger, J., Hengartner, H., & Zinkernagel, R. M. (1991). Quantification of lymphocytic choriomeningitis virus with an immunological focus assay in 24- or 96-well plates. *Journal of Virological Methods*, *33*, 191–198.

Doherty, P. C., & Zinkernagel, R. M. (1975). H-2 compatibility is required for T-cell-mediated lysis of target cells infected with lymphocytic choriomeningitis virus. *The Journal of Experimental Medicine*, *141*, 502–507.

Fooksman, D. R., Vardhana, S., Vasiliver-Shamis, G., Liese, J., Blair, D. A., Waite, J., et al. (2010). Functional anatomy of T cell activation and synapse formation. *Annual Review of Immunology*, *28*, 79–105.

Fung-Leung, W. P., Kundig, T. M., Zinkernagel, R. M., & Mak, T. W. (1991). Immune response against lymphocytic choriomeningitis virus infection in mice without CD8 expression. *The Journal of Experimental Medicine*, *174*, 1425–1429.

Gallimore, A., Glithero, A., Godkin, A., Tissot, A. C., Pluckthun, A., Elliott, T., et al. (1998). Induction and exhaustion of lymphocytic choriomeningitis virus-specific cytotoxic T lymphocytes visualized using soluble tetrameric major histocompatibility complex class I-peptide complexes. *The Journal of Experimental Medicine*, *187*, 1383–1393.

Hangartner, L., Zinkernagel, R. M., & Hengartner, H. (2006). Antiviral antibody responses: The two extremes of a wide spectrum. *Nature Reviews. Immunology*, *6*, 231–243.

Lau, L. L., Jamieson, B. D., Somasundaram, T., & Ahmed, R. (1994). Cytotoxic T-cell memory without antigen. *Nature*, *369*, 648–652.

McGavern, D. B., Christen, U., & Oldstone, M. B. (2002). Molecular anatomy of antigen-specific CD8(+) T cell engagement and synapse formation in vivo. *Nature Immunology, 3*, 918–925.

Murali-Krishna, K., Altman, J. D., Suresh, M., Sourdive, D. J., Zajac, A. J., Miller, J. D., et al. (1998). Counting antigen-specific CD8 T cells: a reevaluation of bystander activation during viral infection. *Immunity, 8*, 177–187.

Ohashi, P. S., Oehen, S., Buerki, K., Pircher, H., Ohashi, C. T., Odermatt, B., et al. (1991). Ablation of "tolerance" and induction of diabetes by virus infection in viral antigen transgenic mice. *Cell, 65*, 305–317.

von Herrath, M. G., Guerder, S., Lewicki, H., Flavell, R. A., & Oldstone, M. B. (1995). Coexpression of B7-1 and viral ("self") transgenes in pancreatic beta cells can break peripheral ignorance and lead to spontaneous autoimmune diabetes. *Immunity, 3*, 727–738.

Welsh, R. M., & Seedhom, M. O. (2008). Lymphocytic choriomeningitis virus (LCMV): Propagation, quantitation, and storage. *Current Protocols in Microbiology, Chapter 15*. Unit 15A 11.

Wherry, E. J., Blattman, J. N., Murali-Krishna, K., van der Most, R., & Ahmed, R. (2003). Viral persistence alters CD8 T-cell immunodominance and tissue distribution and results in distinct stages of functional impairment. *Journal of Virology, 77*, 4911–4927.

Zhou, X., Ramachandran, S., Mann, M., & Popkin, D. L. (2012). Role of lymphocytic choriomeningitis virus (LCMV) in understanding viral immunology: Past, present and future. *Viruses, 4*, 2650–2669.

Zinkernagel, R. M., & Doherty, P. C. (1974). Immunological surveillance against altered self components by sensitised T lymphocytes in lymphocytic choriomeningitis. *Nature, 251*, 547–548.

Zinkernagel, R. M., & Doherty, P. C. (1975). H-2 compatability requirement for T-cell-mediated lysis of target cells infected with lymphocytic choriomeningitis virus. Different cytotoxic T-cell specificities are associated with structures coded for in H-2K or H-2D. *The Journal of Experimental Medicine, 141*, 1427–1436.

Zinkernagel, R. M., Haenseler, E., Leist, T., Cerny, A., Hengartner, H., & Althage, A. (1986). T cell-mediated hepatitis in mice infected with lymphocytic choriomeningitis virus. Liver cell destruction by H-2 class I-restricted virus-specific cytotoxic T cells as a physiological correlate of the 51Cr-release assay? *The Journal of Experimental Medicine, 164*, 1075–1092.

CHAPTER 10

An in vitro model to monitor natural killer cell effector functions against breast cancer cells derived from human tumor tissue

Nicky A. Beelen[a,b,c], Femke A.I. Ehlers[a,b,c], Loes F.S. Kooreman[b,d], Gerard M.J. Bos[a,b], and Lotte Wieten[b,c,*]

[a]Department of Internal Medicine, Division of Hematology, Maastricht University Medical Center, Maastricht, The Netherlands
[b]GROW-School for Oncology and Reproduction, Maastricht University, Maastricht, The Netherlands
[c]Department of Transplantation Immunology, Tissue Typing Laboratory, Maastricht University Medical Center, Maastricht, The Netherlands
[d]Department of Pathology, Maastricht University Medical Center, Maastricht, The Netherlands
*Corresponding author: e-mail address: l.wieten@mumc.nl

Chapter outline

1 Introduction 135
2 Before you begin 137
3 Key resources table 138
4 Materials and equipment 139
5 Step-by-step method details 140
   5.1 Dissociation of breast cancer tissue into single cells and tumor cell enrichment 140
      5.1.1 Tumor sample processing (according to protocol Miltenyi Biotec ("Tumor Dissociation Kit, human")) 140
      5.1.2 Assessment of viability and purity of tumor cells 141
   5.2 Isolation of NK cells from peripheral blood mononuclear cells (PBMCs) to be used for functional assays against human primary breast cancer cells 143
      5.2.1 Isolation of PBMCs 143
      5.2.2 Isolation of NK cells by negative selection from PBMCs (according to protocol Miltenyi Biotec ("NK Cell Isolation Kit, human")) 143

    5.3 NK cell functional assays against human primary breast cancer cells.....144
        5.3.1 NK cell cytotoxicity assay..............144
        5.3.2 Assessment of NK cell secretory profile, i.e. cytokine assays (Optional)..............145
        5.3.3 NK cell degranulation assay, i.e. CD107a assay, against human primary breast cancer cells..............146
6 Expected outcomes..............148
    6.1 Dissociation of breast cancer tissue..............148
7 Quantification and statistical analysis..............148
    7.1 Data analysis NK cell cytotoxicity assay..............148
    7.2 Data analysis NK cell CD107a assay..............148
8 Advantages..............151
9 Limitations..............151
10 Safety considerations and standards..............151
References..............152

Abstract

Adoptive natural killer (NK) cell-based immunotherapy poses a promising treatment approach in cancer. Despite minimal toxicities associated with NK cell infusion, the potential of NK cell therapy is inhibited by the immunosuppressive tumor microenvironment (TME). Multiple approaches to improve anti-cancer NK cell effector functions are being investigated. While much of this preclinical research is currently performed with commercially available tumor cell lines, this approach lacks the influence of the TME and heterogeneity of the primary tumor in patients. Here, we describe a comprehensive protocol for NK cell cytotoxicity- and degranulation assays against tumor cells derived from primary breast cancer tissue. Treatments to boost NK cell anti-tumor effector functions can be implemented in this model. Moreover, by using culture supernatants in follow up assays or by including additional cell types in the co-culture system, other NK cell effector mechanisms that further orchestrate innate and adaptive immunity could be studied.

List of abbreviations

Ac/Dc   acceleration/deceleration
ADCC   antibody-dependent cellular cytotoxicity
E:T ratio   effector-to-target cell ratio
FCS   fetal calf serum
FMO   fluorescence-minus-one
HSA   human serum albumin
NK cell   natural killer cell

PanCK	pancytokeratin
PFA	paraformaldehyde
P/S	penicillin-streptomycin
PBMC	peripheral blood mononuclear cell
PBS	phosphate-buffered saline
RT	room temperature
TME	tumor microenvironment

1 Introduction

Adoptive natural killer (NK) cell-based immunotherapy is a promising treatment approach in oncology. NK cells, part of the innate immune response, are first line responders against tumors and can kill tumor cells without prior immunization. Moreover, NK cells release a plethora of cytokines that influence adaptive immunity (Shimasaki, Jain, & Campana, 2020). The success of NK cell immunotherapy is dependent on the ability of NK cells to exert their anti-tumor effector function within an immune suppressive tumor microenvironment (TME) (Vitale, Cantoni, Pietra, Mingari, & Moretta, 2014). As a consequence of the immunosuppressive TME, the infusion of adoptive NK cells alone, while showing minimal toxicities, does not yet achieve sufficient clinical successes (Miller et al., 2005; Myers & Miller, 2021). However, clinical trials with NK-based therapies in combination with other modalities, such as high dose chemotherapy, activating antibodies, or genetically engineered NK cells, are showing promising initial results and more trials are ongoing (Ciurea et al., 2021; Dolstra et al., 2017; Liu et al., 2021).

Preclinical research into enhancing healthy donor-derived NK cell effector function is currently mainly performed with commercially available tumor cell lines. However, these tumor cell lines have not been under the influence of an in vivo TME, which may directly affect the phenotype of cells from commercial lines. For example, the expression level of human leukocyte antigen (HLA)-E is low on a commercial cell line cultured in vitro, while patient-derived primary cells expressed relatively high levels (Sarkar et al., 2015). Moreover, culturing the same cell line in vivo increased the HLA-E expression level, indicating that some in vivo factors are required to maintain the tumor cells original phenotype (Sarkar et al., 2015). Similar observations are made in regard to other HLA class I molecules and the HLA class I antigen-processing machinery (APM) (Cai et al., 2018). However, the mechanism behind the discrepancies in HLA class I expression between long term established tumor cell lines and the in vivo tumor is not yet understood (Cai et al., 2018; Garrido, 2019). In line with this, commercially available tumor cell lines do not capture the heterogeneity that occurs in primary tumor tissue (Gillet et al., 2011). To better incorporate the effects of the clinically relevant TME on tumor cells and the expected intratumoral-heterogeneity into NK cell research, we established a protocol to use tumor tissue from breast cancer patients in combination with commonly used assays to determine the effector functions of healthy donor-derived

NK cells (Alter, Malenfant, & Altfeld, 2004; Bryceson et al., 2010; Cruz et al., 2005; Ehlers et al., 2021). The use of primary tumor tissue has benefits over long-time passaged 2D cancer cell lines, since physiological characteristics of cell lines may have drifted away from the original tumors, due to genomic instability and influence of varying or non-physiological culture methods (Ben-David et al., 2018; Daniel et al., 2009; Gillet et al., 2011; Landry et al., 2013; Sandberg & Ernberg, 2005). Moreover, these tumor cells have been exposed to the TME and its priming effects may be maintained ex vivo. Indeed, it is reported that dissociated patient-derived material from various tumor types better represents the original tumor epithelial cells compared to cancer cell lines (Ghani et al., 2019; Oppel et al., 2019). Furthermore, the generation of 2D cultures from patient material allows for further experimentation, such as the co-culture with various immune cells to simulate and investigate cell-cell communication, which could lead to new strategies for breast cancer therapy.

New therapeutic options for breast cancer are required as female breast cancer is currently the most commonly diagnosed cancer worldwide and a leading cause of death (Sung et al., 2021). Due to earlier detection by mammography screening and treatment improvements, survival rates have increased over the past decades. However, these improvements are not yet observable for all subtypes, nor for metastatic disease (Miller et al., 2019). As cellular immunotherapy may provide a novel treatment option for patients that not sufficiently respond to current therapies, in vitro models with primary tumor cells can be instrumental to further develop and optimize immunotherapy.

Here, we provide a detailed and integrated experimental protocol that describes a model to study the interaction between NK cells obtained from healthy donors and "TME-exposed" primary tumor cells, in which potential therapeutic agents that could boost NK cell anti-tumor effector functions can be tested. Protocols for both tumor specific cytotoxicity and NK cell degranulation (CD107a) are explained in detail. Additionally, the researcher could assess cytokine secretion in the supernatant of the cytotoxicity assays by commercially available platforms for cytokine analysis (e.g. ELISA, Luminex or Cytometric Bead Array) or, when combined with microscopy, use the dissociated tumor cells as a model to study tumor cell-NK cell interaction at the level of the immunological synapse. This co-culture system is optimized for primary breast cancer material. Other tumor types may be dissociated using a similar protocol, but may behave differently based on their inherent tissue morphology or structure. The protocol is optimized for high yield of tumor cells and many cell surface epitopes recognized by NK cells are preserved ("Tumor Dissociation Kit, human. Preservation of cell surface epitopes"). After the dissociation, tumor cell enrichment is performed, which excludes tumor-infiltrating lymphocytes (TILs) (Fig. 1 & Table 1). However, TILs can be recovered, which would allow for the subsequent culturing of tumor cells and TILs, and the analysis of TIL phenotypes (beyond the scope of this chapter). Other effector mechanisms of NK cells that further orchestrate innate and adaptive immunity could be studied by including these additional cell types in the co-culture system or by using culture supernatants in follow up assays (Bald, Krummel, Smyth, & Barry, 2020).

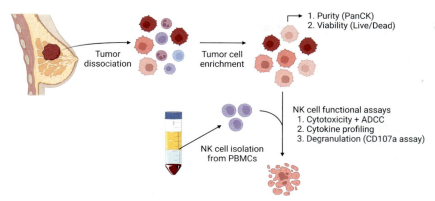

FIG. 1
Schematic overview of methodological approach from tumor tissue until NK cell functional assays. Breast tumor tissue is resected and the solid tumor tissue is dissociated. This results in a cell suspension containing tumor cells and other cells of the TME, including leukocytes. Tumor cell enrichment is performed to exclude other cell compartments. The purity based on PanCytokeratin (PanCK+) and viability (Live/Dead-) of the dissociated breast cancer cells can be assessed with flow cytometry. Enriched tumor cells are co-cultured with NK cells and NK cell functional readouts are performed.

Table 1 Information on the breast cancer tissue, obtained single cells, purity, and viability.

	Mean	Min	Max
Number of total cells after dissociation ($\times 10^6$ cells per gram tissue)	3.7	0.6	10.4
Number of enriched tumor cells ($\times 10^6$ cells per gram tissue)	0.4	0.04	0.9
Enriched tumor cells (% of total dissociated cells)	9	1	25
PanCK positive (PanCK$^+$) cells (% of enriched tumor cells)	77	44	92
Viability of PanCK$^+$ cells after 16h culture (% of PanCK$^+$ cells)	63	48	91

Indication of the obtained yield of single cells after dissociation and after enrichment step for tumor cells. The purity of enriched tumor cells was assessed by flow cytometry staining for PanCytokeratin (PanCK) and spontaneous cell death of PanCK$^+$ tumor cells is shown after 16h incubation. This work was approved by the Institutional Ethics Committee of Maastricht University Medical Center+ (METC 2019–1015, 1 July 2019). (N=8–10 individual samples derived from primary breast cancer tissues).

2 Before you begin

Timing: 5h.
1. Enzymes of tumor dissociation kit
    (a) According to the manufacturer's protocol (Miltenyi Tumor Dissociation Kit, Human (Tumor Dissociation Kit, human, 2022a, 2022b))
2. Isolation buffer for tumor cell enrichment
    (b) PBS/0.5% Human Serum Albumin (HSA)

3. Intracellular staining buffers (according to manufacturer's protocol ("BD Cytofix/Cytoperm™, Fixation/Permeabilization Kit,")) (BD Cytofix/Cytoperm™, 2022)
   (a) Dilute Permeabilization/Wash buffer from $10\times$ to $1\times$ in double distilled (dd) H_2O
   (b) Dilute Fixation/Permeabilization buffer from $4\times$ to $1\times$ using the included diluent
4. Flow cytometry buffer
   (a) PBS/1% Fetal Calf Serum (FCS)/0.02% NaN_3
5. Culture medium
   (a) RPMI/10% FCS/1% P/S
6. EDTA/PBS buffer
   (a) PBS/2 mM EDTA
7. Red blood cell lysis buffer at room temperature (sterile filtered)
   (a) 155 mM NH_4Cl/10 mM $KHCO_3$/1 mM EDTA in MilliQ H_2O
8. MACS buffer (Use MACS buffer cold (4 °C) for optimal results)
   (a) PBS/2 mM EDTA/0.5% HSA
9. Tumor sample acquisition
   (a) Work together with a pathologist to resect tumor material and to accurately separate tumor from healthy tissue. Compared to healthy tissue, tumor tissue will appear more firm, rigid, and solid. Tumor tissue is often white and its invasiveness can often be observed based on the irregular borders.
   (b) After tissue resection, immediately transfer the breast cancer tissue into the MACS tissue storage solution (see table with reagents), which allows storage for 48 h according to the manufacturer's protocol, and store tissues at 4 °C until tissue is processed.

3 Key resources table

Reagent or resource	Source	Identifier
Antibodies		
PanCytokeratin (PanCK), clone C11	Invitrogen	MA5-18156
LIVE/DEAD™ Fixable Aqua Dead Cell Stain Kit, for 405 nm excitation	ThermoFisher	L34966
αCD107a-VioBlue (H4A3)	Miltenyi Biotec	130-119-873
gG1κ-VioBlue (IS5-21F5)	Miltenyi Biotec	130-094-670
αCD56- PerCP-Vio700 (REA196)	Miltenyi Biotec	130-100-681
CD3- APC-Vio770 (SK7)	Miltenyi Biotec	130-096-610
Flow Cytometry Compensation beads	Miltenyi Biotec and BD Biosciences	130-104-693 and 552843

—cont'd

Reagent or resource	Source	Identifier
Chemicals, peptides, and recombinant proteins		
MACS Tissue Storage Solution	Miltenyi Biotec	130-100-008
Phosphate Buffered Saline (PBS) (1 ×)	Sigma	D8537-500ML
Human Serum Albumin (HSA)	e.g. Albuman (200 g/L)	
NaN_3	Sigma-Aldrich	S2002
RPMI-1640 (1 ×) + GlutaMAX (hereafter referred to as RPMI)	Gibco	61,870,036
Fetal Calf Serum (Heat Inactivated) (FCS)	e.g. Sigma Aldrich	F7524
Penicillin-Streptomycin (10,000 U/mL) (P/S)	Gibco	15,140,122
Tumor Dissociation Kit, human	Miltenyi Biotec	130-095-929
Tumor Cell Isolation Kit, human	Miltenyi Biotec	130-108-339
BD Cytofix/Cytoperm™ Fixation/Permeabilization Kit	BD Biosciences	554,714
Lymphoprep	Axis shield	04-03-9391/02
200 mM EDTA	e.g. ThermoFisher	42-500-5004
Human NK isolation kit	Miltenyi	130-092-657
RPMI-1640 (1 ×) + GlutaMAX (hereafter referred to as RPMI)	Gibco	61870036
BD Golgi Stop (Monensin)	BD Biosciences	554724
Software and algorithms		
FlowJo™ Software	BD Biosciences	
BD FACSDiva™ Software	BD Biosciences	

4 Materials and equipment

Name	Company + Cat. No.
gentleMACS Dissociator	Miltenyi Biotec (130-093-235)
gentleMACS C Tubes	Miltenyi Biotec (130-096-334)
100 μm and 70 μm EASYstrainer	Greiner Bio-one (542000 and 542070)
Pre-separation filter	Miltenyi (130-041-407)
LS columns	Miltenyi Biotec (130-042-401)
MACS magnetic cell separator (e.g. MACS MultiStand)	Miltenyi Biotec (130-042-303)
Tube rotator at 37 °C (e.g. MACSmix Tube Rotator)	Miltenyi Biotec (130-090-753)
Gas controlled CO_2 incubator (37 °C, 21% O_2, 5% CO_2)	e.g. Sanyo MCO-20AIC, Sanyo Electric Co, Japan

Continued

—cont'd

Name	Company + Cat. No.
Cooling centrifuge with swing-out rotor	e.g. Rotanta 460 R, Hettich Zentrifugen (5660 with 5624 rotor)
Flow Cytometer with 633 nm (Red), 488 nm (Blue), 405 nm (Violet) laser	e.g. BD FACS Canto II
Conical centrifuge tubes (15 and 50 mL)	e.g. Corning (352096 and 352070)
Micropipettes (0.2–2 μL, 1–20 μL, 20–200 μL, 200–1000 μL) and corresponding tips	e.g. Gilson (F167300)
Automatic pipette and serological pipettes (5, 10, and 25 mL)	e.g. Corning (4099) and Corning (4487, 4488, and 4489)
Hemocytometer (e.g. Bürker counting chamber)	VWR (630-1542)
Heparin tubes e.g. BD vacutainer (if peripheral blood is PBMC source)	BD Biosciences (366480)
96 wells round bottom plate	e.g. Corning (3799)

5 Step-by-step method details

5.1 Dissociation of breast cancer tissue into single cells and tumor cell enrichment

Timing: 4 h.

5.1.1 Tumor sample processing (according to protocol Miltenyi Biotec ("Tumor Dissociation Kit, human"))

1. Work in the flow cabinet to keep the procedure as sterile as possible.
2. Transfer the tissue to a Petri dish containing enough RPMI to prevent the tissue from drying out.
3. Weigh tissue by weighing the Petri dish (with the added RPMI) before and after transferring the tissue to the Petri dish.
    Note: The reagent volumes below are described for up to 1 g of tumor tissue. When more tissue is processed, adjust the volumes accordingly.
4. With a scalpel, cut tumor tissue in pieces, as small as possible.
5. Transfer tissue pieces and RPMI to the C-tube.
6. Rinse the Petri dish with 1–2 mL RPMI and add it to the C-tube.
7. For up to 1 g of tissue, add volume in the C-tube up to 4.7 mL with RPMI and add 200 μL Enzyme H, 100 μL Enzyme R, 25 μL Enzyme A.
8. Tightly close the C-tube and attach it upside down onto the gentle MACS Dissociator.
9. Run program h_tumor_01.
10. Attach the C-tube onto the tube rotator and let it rotate at 37 °C for 30 min (min).
11. Repeat step 8–10.

12. Attach C-tube to gentle MACS Dissociator and run program h_tumor_01.
13. Place a 100 μM cell strainer on a new 50 mL tube and filter the cell suspension.
14. Mash the remaining tumor pieces through the filter, e.g. with the plunger of a syringe.
15. Rinse the C-tube and cell strainer with 5–10 mL RPMI.
16. Place a 70 μM cell strainer on a new 50 mL tube and filter the cell suspension.
17. Centrifuge the cell suspension at $280 \times g$ for 8 min at room temperature (RT) with the highest acceleration (ac = 9) and the highest deceleration (dc = 9).
18. Carefully remove the supernatant (e.g. with a serological pipette).
19. Resuspend in 1 mL of RPMI.
20. Count the cells (e.g. by using a hemocytometer).
21. To continue with the tumor cell enrichment, centrifuge the cell suspension at $280 \times g$ for 8 min at RT (ac = 9, dc = 9).
22. Resuspend up to 1×10^7 total cells in 60 μL of isolation buffer.
23. Add 20 μL of the Non-Tumor Cell Depletion Cocktail A and 20 μL of the Non-Tumor Cell Depletion Cocktail B (hence, per 1×10^7 cells).
24. Mix by vortexing the tube 3 s and incubate for 15 min at 4 °C.
25. Meanwhile, prepare the magnet with columns: Place LS column in the magnetic field of the MACS separator and rinse the column with 3 mL of isolation buffer.
26. After 15 min incubation, adjust volume to 500 μL with isolation buffer.
27. Apply cell suspension onto the column (max. 4×10^7 total cells per column).
28. Collect flow-through containing unlabeled cells, representing the enriched tumor cells.
29. Wash the column two times with 1 mL of isolation buffer and collect unlabeled cells that pass through in the same tube.
30. Optional: To collect the labeled cells, remove column from the separator and place it onto a new tube. Pipette 3 mL of isolation buffer onto the column and immediately flush out the magnetically labeled non-tumor cells by firmly pushing the plunger into the column.
31. Centrifuge cell suspensions at $280 \times g$ for 8 min at RT (ac = 9, dc = 9).
32. Carefully remove supernatant (e.g. with a serological pipette) and resuspend in 0.5 mL or 1 mL medium depending on pellet size.
33. Count cells after isolation using a hemocytometer.

5.1.2 Assessment of viability and purity of tumor cells
34. Plate 20,000 tumor cells in 100 μL in a 96 well round-bottom plate.
    Note: Plate duplicates per condition.
35. Centrifuge the 96 well round-bottom plate for 3 min at $787 \times g$ at 4 °C (ac = 9, dc = 9).
36. Pulse vortex the cell pellets.
37. Add 200 μL PBS to the cells.

38. Centrifuge 96 well plate for 3 min at 787 × g at 4 °C (ac = 9, dc = 9), discard supernatant by flicking the plate and pulse vortex.
39. Dilute Live/Dead Aqua to 1 × with PBS by adding 1 μL of Live/Dead Aqua per 999 μL PBS.
40. Add 25 μL of Live/Dead Aqua 1 × to respective wells.
41. Incubate 30 min at 4 °C.
42. Wash using 200 μL PBS.
43. Centrifuge 96 well plate for 3 min at 787 × g at 4 °C (ac = 9, dc = 9), discard supernatant by flicking the plate and pulse vortex.
44. Add 200 μL 1 × Fixation/Permeabilization buffer.
45. Incubate for 30–60 min at 2–8 °C and protect from light.
46. Centrifuge 96 well plate for 3 min at 787 × g at 4 °C (ac = 9, dc = 9), discard supernatant by flicking the plate and pulse vortex.
47. Add 200 μL 1 × Permeabilization Buffer to each well and centrifuge samples at 787 × g for 3 min at 4 °C (ac = 9, dc = 9). Discard the supernatant by flicking the plate and pulse vortex.
48. Repeat Step 46 once.
49. Add 100 μL of 1 × Permeabilization Buffer.
50. Dilute the PanCK antibody in the Permeabilization Buffer.
51. Without washing, add the recommended amount of directly conjugated antibody for detection of intracellular antigen to the cells and incubate for 30 min at 4 °C. Protect from light.
    Note: We used 25 μL per well of 1:500 diluted PanCK-AF488 antibody.
52. Add 200 μL of 1 × Permeabilization Buffer to each well and centrifuge samples for 3 min at 787 × g at 4 °C (ac = 9, dc = 9).
53. Discard the supernatant by flicking the plate and pulse vortex.
54. Repeat Step 51–52.
55. Resuspend stained cells in 200 μL of flow cytometry buffer.
56. Measure and analyze the samples on a flow cytometer.

Notes
1. The tumor dissociation procedure is based on the protocol belonging to the Miltenyi Tumor Dissociation Kit, Human ("Tumor Dissociation Kit, human,").
2. Work fast and once single cells are obtained, keep cells cool and use pre-cooled solutions.
3. The Epitope Preservation List provides an overview of cell surface markers that may or may not be preserved in this protocol ("Tumor Dissociation Kit, human. Preservation of cell surface epitopes,").
   a. To preserve more epitopes, enzyme R can be left out or reduced in volume. However, this may result in less thorough dissociation and a lower cell yield.
   b. Miltenyi Biotech recommends reducing enzyme R to 20% for analysis of tumor infiltrating leukocytes.
4. Section 5.1.2 can be performed in parallel to Section 5.3.1, if a cytotoxicity assay is performed.

5.2 Isolation of NK cells from peripheral blood mononuclear cells (PBMCs) to be used for functional assays against human primary breast cancer cells

Timing: 5 h.

5.2.1 Isolation of PBMCs

57. Transfer 15 mL Lymphoprep (at room temperature) into 50 mL tubes.
      Note: When a Buffy coat is used as PBMC source: add 50 mL buffy coat to 50 mL PBS to dilute 1:1.
58. Carefully layer 25 mL buffy coat/PBS or whole blood onto the Lymphoprep. Hold 50 mL tube with Lymphoprep diagonally; slowly add the blood without disturbing the Lymphoprep layer.
59. Centrifuge for 20 min at $600 \times g$ at RT and the lowest ac and dc (ac = 1, dc = 1).
60. Transfer the intermediate layer (the PBMCs) (Fig. 1) into new 50 mL tubes. Add max. 15 mL per 50 mL tube.
61. Add up to 50 mL with 2 mM EDTA/PBS.
62. Centrifuge for 8 min at $280 \times g$ at RT (ac = 9, dc = 9).
63. Discard supernatant (e.g. with a serological pipette).
64. Check for the presence of blood aggregates. If blood aggregates are present:
    a. Filter the cell suspension with a 70 μm filter.
    b. Wash filter and add up to 50 mL with 2 mM EDTA/PBS.
    c. Centrifuge for 8 min at $280 \times g$ at RT (ac = 9, dc = 9).
65. Check for the presence of red blood cells.
      Note: The lymphocyte pellet should appear as a white pellet. Red blood cells will turn the white cell pellet red. It is not necessary to perform red blood cell lysis if only a minor part of the cell pellet is red. If red blood cells are present:
    a. Re-suspend each cell pellet in 5–10 mL lysis buffer (lysis buffer should be at RT).
    b. Pool all cell pellets into one 50 mL tube.
    c. Add lysis buffer up to 30 mL.
    d. Centrifuge for 3 min at $787 \times g$ at RT (ac = 9, dc = 9).
    e. Discard supernatant (e.g. with a serological pipette).
    f. Re-suspend cell pellet (white) in 10 mL 2 mM EDTA/PBS and fill up to 50 mL with 2 mM EDTA/PBS.
    g. Centrifuge for 8 min at $280 \times g$ at RT (ac = 9, dc = 9).
    h. Discard supernatant (e.g. with a serological pipette).
66. Re-suspend pellet in MACS buffer and count the cells (e.g. by using a hemocytometer).

5.2.2 Isolation of NK cells by negative selection from PBMCs (according to protocol Miltenyi Biotec ("NK Cell Isolation Kit, human")

67. Add MACS buffer to make up to 50 mL.
68. Centrifuge for 8 min at $280 \times g$ at RT (ac = 9, dc = 9).

69. Re-suspend in 40 μL cold MACS buffer per 10^7 cells.
70. Store for 5 min at 4 °C.
71. Add 10 μL antibody-cocktail to every 1×10^7 cells.
72. Mix by vortexing the tube 3 s.
73. Incubate for 5 min at 4 °C.
74. Add 20 μL α-biotin beads and 30 μL MACS buffer per 1×10^7 cells.
75. Mix by vortexing the tube 3 s.
76. Incubate for 10 min at 4 °C.
77. Wash the cells including beads in MACS buffer by adding 10 to 20 times the reaction volume.
78. Centrifuge for 8 min at $280 \times g$ at RT (ac = 9, dc = 9).
79. Meanwhile, put a separation filter on top of the LS column and rinse the column with 3 mL MACS buffer.
80. Re-suspend the cells after centrifugation in 500 μL MACS buffer/LS column/ 1×10^8 cells.
81. Collect the run through (NK cells) in a clean tube.
82. Pipette 500 μL of the cells on top of the filter (max 1×10^8 cells per LS column).
83. Rinse the column three times with 3 mL MACS buffer.
84. Centrifuge for 8 min at $280 \times g$ at RT (ac = 9, dc = 9).
85. Discard supernatant (e.g. with a serological pipette).
86. Re-suspend cells in culture medium.
87. Count the cells (e.g. by using a hemocytometer).
88. Use the NK cells in functional assays (see Part 5.3) or activate the NK cells by overnight (>16 h) culture at $1-2 \times 10^6$ cells per mL in culture medium with 500–1000 U/mL IL2 in an incubator at 37 °C with 21% O_2 and 5% CO_2.

Notes
1. The NK cell isolation procedure is based on the protocol belonging to the Miltenyi Human NK cell isolation Kit ("NK Cell Isolation Kit, human, 2022").
2. The purity of the isolated NK cells can be assessed by flow cytometry. NK cells can be identified as being positive for CD56 and negative for CD3. Double positive cells ($CD56^+CD3^+$) indicate NKT cells. Suggested antibodies to use: CD56-APC-Vio770 (Miltenyi; 130-114-739) and CD3-PE-Vio770 (Miltenyi; 130-113-702).

5.3 NK cell functional assays against human primary breast cancer cells

5.3.1 NK cell cytotoxicity assay
Timing: 4 h + 16 h incubation.

89. Plate 20,000 tumor cells in 100 μL in a 96 well round-bottom plate.
    a. Plate duplicates per condition.
    b. Plate 20,000 tumor cells in 100 μL with 100 μL culture medium as control for spontaneous tumor cell death.

90. To prepare NK cells at the desired E:T ratios, adjust the concentration of the NK cells using culture medium as follows:

E:T ratio	1:1	5:1	10:1
Concentration of NK cells	0.2×10^6 cells/mL	1×10^6 cells/mL	2×10^6 cells/mL

91. Transfer 100 μL of NK cells at different concentrations to the respective wells with tumor cells.
92. Incubate cells for 16 h in an incubator at 37 °C with 21% O_2 and 5% CO_2.
93. After incubation put plates on ice.
94. Centrifuge the 96 well round-bottom plate for 3 min at $787 \times g$ at 4 °C (ac = 9, dc = 9), discard supernatant by flicking the plate and pulse vortex.
95. Add 200 μL PBS to the cells.
96. Centrifuge 96 well plate for 3 min at $787 \times g$ at 4 °C (ac = 9, dc = 9), discard supernatant by flicking the plate and pulse vortex.
97. Dilute Live/Dead Aqua to 1× with PBS by adding 1 μL of Live/Dead Aqua per 999 μL PBS.
98. Add 25 μL of 1× Live/Dead Aqua to respective wells.
99. Incubate 30 min at 4 °C.
100. Wash using 200 μL PBS.
101. Centrifuge 96 well plate for 3 min at $787 \times g$ at 4 °C (ac = 9, dc = 9), discard supernatant by flicking the plate and pulse vortex.
102. For the intracellular staining of PanCK, repeat Step 43–54.
103. Resuspend stained cells in 200 μL of flow cytometry buffer.
104. Measure and analyze the samples on a flow cytometer.

    Note: Cells can be fixed by resuspending cells in 200 μL 1% PFA, which allows for flow cytometry analysis at a later time point (maximal 4 days after flow cytometry staining).

5.3.2 Assessment of NK cell secretory profile, i.e. cytokine assays (Optional)

- The supernatant after the NK cell cytotoxicity assay can be harvested (by centrifuging 96 well plate for 3 min at $787 \times g$ at 4 °C (ac = 9, dc = 9) and gently pipetting of supernatant without disturbing cell pellet) and stored at −80 °C. Several cytokine assays (ELISA, Luminex, Cytokine Bead Array) are commercially available and can be implemented to examine the secretory profile of NK cells upon tumor cell exposure.

Notes
1. Cytokine assays cannot be performed in combination with the CD107a assay due to the inclusion of Golgi Stop, which contains Monensin that blocks intracellular protein transport processes in the NK cell.

5.3.3 NK cell degranulation assay, i.e. CD107a assay, against human primary breast cancer cells

Timing: 4 h + 4 h incubation.

105. Adjust both NK cell and tumor cell concentrations to 1×10^6 cells/mL using culture medium.
106. Plate 100 μL of the NK cell suspension per well (= 1×10^5 cells) according to the experimental setup.
107. Add 100 μL of tumor cell suspension per well (= 1×10^5 cells) according to the experimental setup.
108. Prepare CD107a or isotype control antibody mixes in medium for all conditions.
    a. Per well: 5 μL CD107a Horizon V450 antibody + 20 μL medium.
    b. Per well: 0.5 μL IgG1κ V450 + 24.5 μL medium.
109. Directly add 25 μL of CD107a antibody mix or isotype control mix to the NK-tumor suspensions (according to the experimental setup).
110. Dilute BD Golgi Stop 10× by adding 1 μL per 9 μL culture medium.
111. Incubate for 1 h at 37 °C.
112. After 1 h add 6 μL/well 10× diluted solution of BD Golgi Stop and incubate for additional 3 h at 37 °C.
113. After incubation put plate on ice.
114. Pre-cool centrifuge to 4 °C and proceed with flow cytometry staining on ice.
115. Transfer cells to 96 deep well plate for flow cytometry staining.
116. Add 200 μL PBS to the cells in the 96 deep well plate.
117. Centrifuge 96 well plate for 3 min at 787 × g at 4 °C (ac = 9, dc = 9).
118. Gently remove supernatant.
119. Dilute Live/Dead Aqua to 1× with PBS by adding 1 μL of Live/Dead Aqua per 999 μL PBS.
120. Add 25 μL of 1× Live/Dead Aqua to respective wells.
121. Incubate 30 min at 4 °C.
122. Wash using 400 μL PBS.
123. Centrifuge 96 well plate for 3 min at 787 × g at 4 °C (ac = 9, dc = 9).
124. Gently remove supernatant.
125. Add 50 μL antibody mix (Table 2) to all wells.
126. Incubate 30 min at 4 °C.
127. Wash using 400 μL PBS.
128. Centrifuge 96 well plate for 3 min at 787 × g at 4 °C (ac = 9, dc = 9).
129. Gently remove supernatant.

Table 2 Antibody mix for CD107a assay.

Product	μL/staining	Clone	Isotype	Conjugate
CD3	0.5	SK7	IgG2a	APC-Vio770
CD56	0.5	REA196	IgG1	PerCP-Vio700
PBS	49	x	x	x

130. Resuspend stained cells in 200 μL of flow cytometry buffer.
   Note: Cells can be fixed by resuspending cells in 200 μL 1% PFA, which allows for flow cytometry analysis at a later time point (maximal 4 days after flow cytometry staining).
131. Analyze by flow cytometer.

Notes
   Cytotoxicity assay
1. We recommend including a positive control for the NK cell killing potential. The positive control should be a tumor cell line that is easily killed by NK cells, e.g. K562.
2. For the compensation settings of the flow cytometer, plate the following controls:
   a. One well with tumor cells as unstained control.
   b. One well with tumor cells and stained for the Live/Dead marker only.
   c. Two wells with tumor cells and stained for PanCK only (one for compensation controls, one to measure as a sample to facilitate setting the Live/Dead gate).
3. The cytotoxicity assay can be combined with a monoclonal antibody that can trigger antibody-dependent cytotoxicity (ADCC) by pre-incubating the tumor cells with the required concentration of the antibody. E.g. trastuzumab can be used at a final concentration of 1 μg/mL (Ehlers et al., 2021).
   a. Incubate 50 μL tumor cells with 50 μL antibody for 30 min.
   b. After 30 min pre-incubation and without additional washing, 100 μL/well of NK cells, diluted at the required concentration, can be added to the tumor cell-antibody cultures.
4. The duration of the cytotoxicity assay may be prolonged to proceed after 16 h of co-culture.
5. The tumor cells are identified as PanCK$^+$ cells and the Live/Dead marker is used to identify the percentage of dead tumor cells.
6. Optionally, the NK cells can be identified by staining for NKp46 after co-culture or by labeling the NK cells with a dye (e.g. CM-DiI from ThermoFisher) BEFORE the co-culture with tumor cells.
7. The antibodies in these assays are compatible with the fixation and permeabilization steps. The compatibility of other antibodies should be tested before use.
   Degranulation (CD107a) assay
8. We recommend including a positive control for the NK cell degranulation potential. The positive control should be a tumor cell line that easily stimulates NK cells to degranulate, e.g. K562.
9. For the compensation settings of the flow cytometer, plate the following controls:
   a. One well with NK cells as unstained control.
   b. One well with NK cells (exposed to 65 °C for 2 min) and stained for Live/Dead Aqua-V500 only.
   c. One well with NK cells and stained for CD56-PerCP-Vio700 only.

d. One well with compensation beads stained for CD3-APC-Vio770.
   e. One well with compensation beads stained for CD107a-V450.
10. Similar to the cytotoxicity assay, the degranulation assay can be combined with a monoclonal antibody that can trigger ADCC.
11. Living NK cells are identified as CD56$^+$CD3$^-$ Live/Dead$^-$ and for this population the percentage CD107a$^+$ cells is quantified.
12. Additional stainings for other NK cell markers of interest (e.g. KIRs or NKG2A) can be included in the flow cytometry staining steps of the CD107a assay.

6 Expected outcomes
6.1 Dissociation of breast cancer tissue

We observed a yield between 0.5×10^6 cells and 8.9×10^6 cells per gram of tumor tissue. From these cells, between 1% and 25% were recovered as tumor cells after the tumor cell-enrichment steps. Moreover, we further assessed the purity of this enriched tumor cell population by staining with PanCK. The purity of the enriched tumor cells was between 44% and 92% and the viability after 16 h culture was between 48% and 91% (Table 1).

7 Quantification and statistical analysis
7.1 Data analysis NK cell cytotoxicity assay

NK cell mediated cytotoxicity is calculated as the percentage Live/Dead$^+$ tumor cells (PanCK$^+$). To calculate the specific cytotoxicity, correct for spontaneous tumor cell death.

- Dead tumor cells can be gated as shown in the gating strategy (Fig. 2).
- Specific cytotoxicity is calculated as followed:

$$\%\text{specific lysis} = \frac{(\%\text{dead cells in sample} - \text{average \% of spontaneous dead})}{(100 - \text{average of \% spontaneous dead})} * 100\%$$

7.2 Data analysis NK cell CD107a assay

NK cell degranulation is calculated as the percentage CD107a$^+$ cells of total NK cells (CD56$^+$CD3$^-$).

- CD107a$^+$ NK cells can be gated as shown in the gating strategy (Fig. 3).
- NK cell degranulation upon tumor exposure should be corrected for the spontaneous degranulation (as calculated in "NK only" condition).

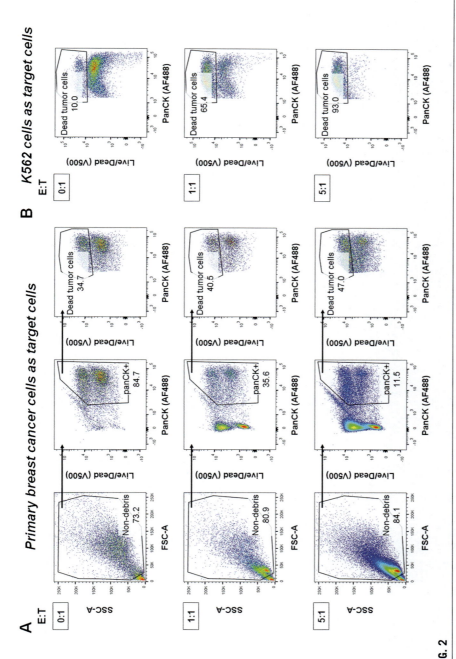

FIG. 2

Gating strategy for cytotoxicity assay. Examples are given for (A) a primary breast cancer sample and (B) the K562 cell line. The upper row shows the condition with tumor cells only, the middle row with tumor cells and NK cells in a 1:1 effector: target (E:T) ratio, and the lower row a 5:1 E:T ratio. From left to right, debris is gated out. Next, the tumor cells are identified by gating on the PanCytokeratin$^+$ (PanCK$^+$) cells. Tumor cell death (Live/Dead$^+$ PanCK$^+$ cells) is determined as percentage of PanCK$^+$ cells. A fluorescence-minus-one (FMO) control sample that is stained for PanCK but not for the LiveDead marker should be included to accurately place the gate to select dead tumor cells.

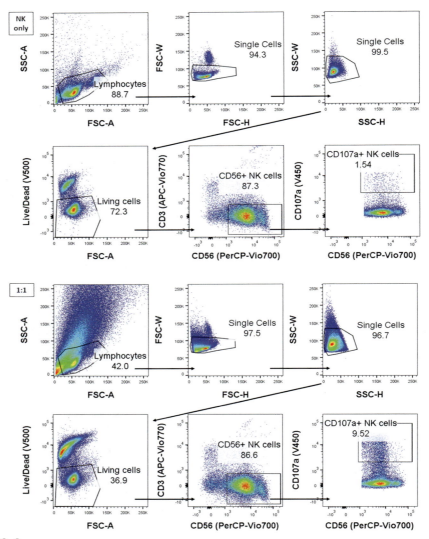

FIG. 3

Gating strategy for degranulation (CD107a) assay. The upper panel shows an example with NK cells only, the lower panel with tumor cells and NK cells in a 1:1 effector: target (E:T) ratio. First, a gate is placed around the lymphocytes. Then doublets and dead cells are gated out. Next, the NK cells are identified by gating on the CD56$^+$CD3$^-$ cells. NK cell degranulation is determined as percentage of CD107a$^+$ NK cells. A fluorescence-minus-one (FMO) control sample that is stained for Live/Dead Aqua but not for the NK cell marker should be included to accurately place the gate to select NK cells. An isotype control for the CD107a antibody can be used to accurately gate on CD107a$^+$ NK cells.

8 Advantages

This protocol describes a model in which a low number of primary breast cancer cells can be used to study multiple NK cell effector functions, i.e. (antibody-dependent cellular-) cytotoxicity, degranulation and cytokine secretion. Moreover, this primary breast cancer model can be successfully combined with standard NK cell functional assays (Alter et al., 2004; Bryceson et al., 2010; Cruz et al., 2005; Ehlers et al., 2021).

9 Limitations

Although the use of patient-derived material better recapitulates the genetic heterogeneity of the original tumor and incorporates the potential effects that the TME may have had in vivo on the tumor cells (e.g. altered surface expression of NK cell ligands), it is not possible to conserve all potential NK cell inhibitory effects of the TME in this ex vivo cytotoxicity assay. Examples for the latter are tumor-derived soluble factors or the influence of the surrounding tumor stroma. However, the primary breast cancer cells have been primed by the TME in the patients, whereas other factors, such as hypoxia or lack of nutrients could be included in this model, such as culturing cells in hypoxic chamber or with physiological culture media, respectively.

We observed variation in viability of the primary tumor cells after 16-h culture due to the assay procedure. The variation in primary tumor cell viability does result in variation in spontaneous tumor cell death during the NK cell functional assays. The correction for spontaneous death is therefore always required when comparing data from different isolations from different patients.

Moreover, the single cell state of these patient tissues does not have to be the only endpoint. Current advances allow for the generation of 3D organoid structures, which are a better representation of the original tumor morphological structure, hence the tumor biology is better maintained compared to dissociated 2D material, and especially compared to commercial cell lines (Yuki, Cheng, Nakano, & Kuo, 2020). The production and utilization of ex vivo patient material has important potential for personalized medicine. As treatments may only be effective for a subset of patients due to heterogeneity, testing potential combinations of immunotherapy and other therapeutic means (e.g. monoclonal antibodies, bispecific antibodies, immunomodulatory agents, or radiotherapy) in this model, will aid in treatment decision making and ultimately lead to personalized medicine that is specific, targeted and effective in treating that patient (Kodack et al., 2017).

10 Safety considerations and standards

As this work includes the handling of blood in the lab, the acting scientists should be vaccinated against hepatitis B.

References

Alter, G., Malenfant, J. M., & Altfeld, M. (2004). CD107a as a functional marker for the identification of natural killer cell activity. *Journal of Immunological Methods*, *294*(1–2), 15–22.

Bald, T., Krummel, M. F., Smyth, M. J., & Barry, K. C. (2020). The NK cell–cancer cycle: Advances and new challenges in NK cell–based immunotherapies. *Nature Immunology*, *21*(8), 835–847.

BD Cytofix/Cytoperm™. (2022). *Fixation/permeabilization kit*. Retrieved from https://www.bdbiosciences.com/content/dam/bdb/products/global/reagents/flow-cytometry-reagents/research-reagents/buffers-and-supporting-reagents-ruo/554714_base/pdf/554714_554715_555028_Book_Website.pdf.

Ben-David, U., Siranosian, B., Ha, G., Tang, H., Oren, Y., Hinohara, K., et al. (2018). Genetic and transcriptional evolution alters cancer cell line drug response. *Nature*, *560*(7718), 325–330.

Bryceson, Y. T., Fauriat, C., Nunes, J. M., Wood, S. M., Björkström, N. K., Long, E. O., et al. (2010). Functional analysis of human NK cells by flow cytometry. *Natural Killer Cell Protocols*, 335–352. Springer.

Cai, L., Michelakos, T., Yamada, T., Fan, S., Wang, X., Schwab, J. H., et al. (2018). Defective HLA class I antigen processing machinery in cancer. *Cancer Immunology, Immunotherapy*, *67*(6), 999–1009.

Ciurea, S. O., Kongtim, P., Soebbing, D., Trikha, P., Behbehani, G., Rondon, G., et al. (2021). Decrease post-transplant relapse using donor-derived expanded NK-cells. *Leukemia*, 1–10.

Cruz, I., Ciudad, J., Cruz, J. J., Ramos, M., Gómez-Alonso, A., Adansa, J. C., et al. (2005). Evaluation of multiparameter flow cytometry for the detection of breast cancer tumor cells in blood samples. *American Journal of Clinical Pathology*, *123*(1), 66–74.

Daniel, V. C., Marchionni, L., Hierman, J. S., Rhodes, J. T., Devereux, W. L., Rudin, C. M., et al. (2009). A primary xenograft model of small-cell lung cancer reveals irreversible changes in gene expression imposed by culture in vitro. *Cancer Research*, *69*(8), 3364–3373.

Dolstra, H., Roeven, M. W., Spanholtz, J., Hangalapura, B. N., Tordoir, M., Maas, F., et al. (2017). Successful transfer of umbilical cord blood CD34+ hematopoietic stem and progenitor-derived NK cells in older acute myeloid leukemia patients. *Clinical Cancer Research*, *23*(15), 4107–4118.

Ehlers, F. A., Beelen, N. A., van Gelder, M., Evers, T. M., Smidt, M. L., Kooreman, L. F., et al. (2021). ADCC-inducing antibody trastuzumab and selection of KIR-HLA ligand mismatched donors enhance the NK Cell anti-breast Cancer response. *Cancers*, *13*(13), 3232.

Garrido, F. (2019). MHC/HLA class I loss in cancer cells. *MHC Class-I Loss and Cancer Immune Escape*, 15–78.

Ghani, F. I., Dendo, K., Watanabe, R., Yamada, K., Yoshimatsu, Y., Yugawa, T., et al. (2019). An ex-vivo culture system of ovarian cancer faithfully recapitulating the pathological features of primary tumors. *Cell*, *8*(7), 644.

Gillet, J.-P., Calcagno, A. M., Varma, S., Marino, M., Green, L. J., Vora, M. I., et al. (2011). Redefining the relevance of established cancer cell lines to the study of mechanisms of clinical anti-cancer drug resistance. *Proceedings of the National Academy of Sciences of the United States of America*, *108*(46), 18708–18713.

References

Kodack, D. P., Farago, A. F., Dastur, A., Held, M. A., Dardaei, L., Friboulet, L., et al. (2017). Primary patient-derived cancer cells and their potential for personalized cancer patient care. *Cell Reports*, *21*(11), 3298–3309.

Landry, J. J., Pyl, P. T., Rausch, T., Zichner, T., Tekkedil, M. M., Stütz, A. M., et al. (2013). The genomic and transcriptomic landscape of a HeLa cell line. *G3: Genes, Genomes Genetics*, *3*(8), 1213–1224.

Liu, S., Galat, V., Galat, Y., Lee, Y. K. A., Wainwright, D., & Wu, J. (2021). NK cell-based cancer immunotherapy: From basic biology to clinical development. *Journal of Hematology & Oncology*, *14*(1), 1–17.

Miller, K. D., Nogueira, L., Mariotto, A. B., Rowland, J. H., Yabroff, K. R., Alfano, C. M., et al. (2019). Cancer treatment and survivorship statistics, 2019. *CA: a Cancer Journal for Clinicians*, *69*(5), 363–385.

Miller, J. S., Soignier, Y., Panoskaltsis-Mortari, A., McNearney, S. A., Yun, G. H., Fautsch, S. K., et al. (2005). Successful adoptive transfer and in vivo expansion of human haploidentical NK cells in patients with cancer. *Blood*, *105*(8), 3051–3057.

Myers, J. A., & Miller, J. S. (2021). Exploring the NK cell platform for cancer immunotherapy. *Nature Reviews Clinical Oncology*, *18*(2), 85–100.

NK Cell Isolation Kit, Human. (2022). Retrieved from https://www.miltenyibiotec.com/upload/assets/IM0001512.PDF.

Oppel, F., Shao, S., Schürmann, M., Goon, P., Albers, A. E., & Sudhoff, H. (2019). An effective primary head and neck squamous cell carcinoma in vitro model. *Cell*, *8*(6), 555.

Sandberg, R., & Ernberg, I. (2005). Assessment of tumor characteristic gene expression in cell lines using a tissue similarity index (TSI). *Proceedings of the National Academy of Sciences of the United States of America*, *102*(6), 2052–2057.

Sarkar, S., van Gelder, M., Noort, W., Xu, Y., Rouschop, K., Groen, R., et al. (2015). Optimal selection of natural killer cells to kill myeloma: The role of HLA-E and NKG2A. *Cancer Immunology, Immunotherapy*, *64*(8), 951–963.

Shimasaki, N., Jain, A., & Campana, D. (2020). NK cells for cancer immunotherapy. *Nature Reviews Drug Discovery*, *19*(3), 200–218.

Sung, H., Ferlay, J., Siegel, R. L., Laversanne, M., Soerjomataram, I., Jemal, A., et al. (2021). Global cancer statistics 2020: GLOBOCAN estimates of incidence and mortality worldwide for 36 cancers in 185 countries. *CA: a Cancer Journal for Clinicians*, *71*(3), 209–249.

Tumor Dissociation Kit, human. (2022a). Retrieved from https://www.miltenyibiotec.com/upload/assets/IM0002061.PDF.

Tumor Dissociation Kit, human. (2022b). *Preservation of cell surface epitopes*. Retrieved from https://www.miltenyibiotec.com/_Resources/Persistent/799525cfa5dbf3781cbe4af9a6d8f2fd1ef8b088/Epitope%20preservation%20list%20TDK%2C%20human.pdf.

Vitale, M., Cantoni, C., Pietra, G., Mingari, M. C., & Moretta, L. (2014). Effect of tumor cells and tumor microenvironment on NK-cell function. *European Journal of Immunology*, *44*(6), 1582–1592.

Yuki, K., Cheng, N., Nakano, M., & Kuo, C. J. (2020). Organoid models of tumor immunology. *Trends in Immunology*, *41*(8), 652–664.

CHAPTER 11

Standardized protocol for the evaluation of chimeric antigen receptor (CAR)-modified cell immunological synapse quality using the glass-supported planar lipid bilayer

Jong Hyun Cho[a,b,†], Wei-chung Tsao[a,b,†], Alireza Naghizadeh[a,b], and Dongfang Liu[a,b,*]

[a]Department of Pathology, Immunology and Laboratory Medicine, Rutgers University-New Jersey Medical School, Newark, NJ, United States
[b]Center for Immunity and Inflammation, New Jersey Medical School, Rutgers-The State University of New Jersey, Newark, NJ, United States
*Corresponding author: e-mail address: dongfang.liu@rutgers.edu

Chapter outline

1. Introduction 156
2. Before you begin 158
3. Key resources table 159
4. Materials and equipment 160
   - 4.1 Chemicals, resources, and reagents 160
   - 4.2 Equipment and software 161
5. Step-by-step method details 161
   - 5.1 Liposome preparation 161
   - 5.2 Dialysis of liposomes 162
   - 5.3 Procedure for immunofluorescence of cells on lipid bilayers in flow chamber slides 162

[†]These authors contributed equally to this work.

5.4 Immunofluorescence imaging using semi-automated image capture and analysis..166
   5.5 Machine learning-based CAR IS quality analysis......................................167
6 **Expected outcomes**..**167**
7 **Statistical analysis**...**169**
8 **Limitations**...**169**
References...**170**

Abstract

Chimeric antigen receptor (CAR)-modified cell therapy is an effective therapy that harnesses the power of the human immune system by re-engineering immune cells that specifically kill tumor cells with tumor antigen specificity. Key to the effective elimination of tumor cells is the establishment of the immunological synapse (IS) between CAR-modified immune cells and their susceptible tumors. For functional activity, CAR-modified cells must form an effective IS to kill tumor cells specifically. The formation of the CAR-specific IS requires the coordination of many cellular processes including reorganization of the cytoskeletal structure, polarization of lytic granules, accumulation of tumor antigen, and phosphorylation of key signaling molecules within the IS. Visualization and assessment of the CAR IS quality can reveal much about the molecular mechanisms that underlie the efficacy of various CAR-modified immune cells. This chapter provides a standardized method of assessing the IS quality by quantifying the tumor antigen (defining the CAR IS formation), cytoskeleton (key component of CAR IS structure), and various molecules of interest involved in the IS formation (key molecular mechanism signatures of CAR IS function) using immunofluorescence on the glass-supported planar lipid bilayer, with a focus on tumor antigen only in this study. We provide specific insights and helpful tips for reagent and sample preparation, assay design, and machine learning (ML)-based data analysis. The protocol described in this chapter will provide a valuable tool to visualize and assess the IS quality of various CAR-modified immune cells.

Abbreviations

CAR   chimeric antigen receptor
IS   immunological synapse
ML   machine-learning
NK   natural killer

1 Introduction

The immunological synapse is the tightly regulated interface between an immune cell and its target cell (e.g., antigen-presenting cells). The IS is formed by two separate cells such as lymphocytic cells, including T-cells, B-cells, natural killer (NK) cells, and chimeric antigen receptor (CAR)-modified cells and their interacting cells. The coordination of IS formation involves several functional steps, including the recognition of antigen, clustering of antigen and other adhesion or costimulatory molecules, activation of intracellular signaling, and polarization of lytic granules. The formation of TCR-mediated IS has been extensively studied (Dustin & Long, 2010; Graf, Bushnell, & Miller, 2007). However, much remains unclear about the IS formation of CAR-modified cells and how it contributes to the efficacy in vitro and in vivo.

Recently, evidence has suggested that higher IS quality correlates with increased efficacy of CAR-modified cells in clinical outcomes (Naghizadeh et al., 2022). In a proof-of-concept paper, 2nd generation (CD3ζ) and 3rd generation (CD3ζ+4-1BB) CAR constructs were compared using traditional cytotoxicity studies and IS quality assessment. The third generation of CAR-T IS quality repeatedly had higher fluorescence intensities of F-actin, Lck, as well as phospho-CD3ζ, and ZAP70. However, standard chromium release assay was unable to differentiate the 2nd and 3rd generation CAR. Moreover, IS quality assessment was applied and accurately predicted the increased efficacy of newly generated CAR-modified cells (Halim et al., 2022; Singh et al., 2021; Xiong et al., 2018).

Current common methods of analyzing the IS include immunofluorescence of cell-cell conjugates, flow cytometry, antigen coating on wells, and glass-supported planar lipid bilayers. IS assessed by cell-cell conjugates is two-dimensional and usually restricts the visibility of the IS structure due to the limits on its orientation (Monks, Freiberg, Kupfer, Sciaky, & Kupfer, 1998). Using flow cytometry systems such as the Amnis ImageStream^x Mk II, cell-cell conjugates can be stained, counted, and quantified to assess IS formation (Markey, Gartlan, Kuns, MacDonald, & Hill, 2015). Although flow cytometry makes IS quantification easier, the resolution of the cell-cell conjugates can limit the assessment of various parameters that contribute to the IS formation. For a "top-down" view of the immunological synapse, antigen coating using stimulation ligands on culture plates has been commonly used to study B-cell receptor (BCR) or T-cell receptor (TCR) immunological synapses (Obino et al., 2017; Purbhoo et al., 2010). Although the orientation makes the visualization of the IS structure clearer, the antigen crosslinked to the plate is unable to move laterally which may hinder IS formation or give inaccurate results on the IS quality. Because of this, the glass-supported lipid bilayer assay has been more meaningful in studying IS structure and quality (Grakoui et al., 1999; Liu, Peterson, & Long, 2012; Majzner et al., 2020). The lipid bilayers offer benefits in that they can mimic the fluidity of the plasma membrane and allow free diffusion of proteins in the bilayer. This enables the rearrangement of proteins upon interaction with CAR-modified cells, which can be imaged at high resolution to study the accumulation of molecules to the synapse such as tumor antigen. Additionally, the supported lipid bilayer system can serve as a reductionist model of the plasma membrane to evaluate the contribution of individual receptors and ligands (Dustin, Starr, Varma, & Thomas, 2007; Liu et al., 2020), which can be developed into a variety of potential high-throughput methods. Finally, the difference on images between real CAR IS formation and un-specific physical interaction can be distinguished by the accumulation of tumor antigen on the glass-supported planar lipid bilayer.

Here, we present a protocol to study the IS using immunofluorescence on the glass-supported planar lipid bilayers. We provide tips and alternative methods within the protocol to obtain a range of antigen concentrations ideal for assessing the IS quality of specific CAR-modified cells. Overall, we provide a standardized method for using the lipid bilayers to analyze known molecules of interest (F-actin, perforin, CD3ζ) which can be adapted to interrogate new antigen or intracellular proteins of interest.

2 Before you begin

Timing: 1–2 days

1. Prepare lipid resuspension and dialysis buffer.
    a. Lipid dialysis buffer – Make 5 L of 25 mM Tris pH 8.0, 150 mM NaCl.
        i. Degas and parafilm seal, store at 4 °C.
    b. Lipid resuspension buffer – 25 mM Tris pH 8.0, 150 mM NaCl, 2% n-octyl-β-D-glucopyranoside (OG).
        i. Around 10 mL per vial of lipids.
        ii. To make 2% OG for DOPC and Biotin-PE, add 0.40 g of OG into 20 mL of lipid dialysis buffer.
2. Clean glass chromatography tubes to store lipids.
    a. Prepare cleaning solution.
        i. Dissolve 60 g of potassium hydroxide (KOH) in 120 mL of H_2O and 1 L 95% ethanol.
    b. Submerge tubes in the cleaning solution for 30 min. Rinse with diH_2O and autoclave. Use caution and tongs when handling tubes as the cleaning solution can break through gloves.
3. Prepare 5% casein.
    a. Dissolve 18 g of casein in 250 mL of ultrapure water while stirring at room temperature (RT) for at least 2–3 h and slowly stir overnight at 4 °C.
    b. Add 36 mL of 10 × PBS and bring volume to 360 mL with ultrapure water. pH should be about 7.3 after adding PBS.
    c. Spin casein in an ultracentrifuge at 40,000 rpm (>100,000 g) from 1.5 to 2 h.
    d. After centrifuge, the tubes will have a precipitate on the bottom and an opaque cloudy layer on top. Carefully remove only the clear solution in the middle layer and place it in a new 50 mL conical tube.
    e. Filter casein with 0.2 μm membrane.
        i. Optional: If the solution is difficult to filter, add collected casein to the new ultracentrifuge at 40,000 rpm (>100,000 g) for an additional 2 h.
        ii. Collect the middle layer and filter.
        iii. Aliquot and freeze at −80 °C standing upright.
4. Prepare HEPES Buffered Saline (HBS) wash buffer.
    a. 20 mM HEPES, 150 mM NaCl, 1% Human serum albumin (HSA).
5. Prepare fluorescence-conjugated proteins.
    a. Proteins conjugated with biotin and His-tags can be used to embed proteins onto the lipid bilayer.
    b. Proteins with biotin or His-tags can be further conjugated with fluorescent dyes such as the Alexa Fluor dyes.
    c. Different CAR target binding sites and reactive sites from the conjugation kits with the tags need to be evaluated and optimized to provide consistent data with different CAR molecules and antigens.

6. Prepare permeabilization buffer.
   a. 10% Normal donkey serum, 0.5% Triton-X in PBS.
7. Prepare antibody-dilution buffer for primary and secondary antibody staining.
   a. 3% Normal donkey serum, 0.05% Triton-X in PBS.

3 Key resources table

Note that not all areas will be used in every protocol.

Reagent or resource	Source	Identifier
Antibodies		
Alexa-Fluor 405 Phalloidin	Thermofisher	A30104
Alexa-Fluor 568 mouse anti-perforin (dG9)	Biolegend	308102
Alexa Fluor 647 rabbit anti-CD3 zeta (phosphor Y83)	Abcam	Ab68236
Chemicals, Peptides, and Recombinant Proteins		
18:1(Δ-Cis) PC (DOPC) 1,2-dioleoyl-sn-glycero-3-phosphocholine (10 mg/mL)	Avanti Polar Lipids	850375C
18:1 Biotinyl Cap PE 1,2-dioleoylsn-glycero-3-phosphoethanolamine-N-(cap biotinyl) (10 mg/mL)	Avanti Polar Lipids	870273C
Compressed Argon gas	Airgas	AR UHP180LT230
Sulfuric Acid	Acros Organics	7664-93-9
Hydrogen peroxide	Sigma	386,90
Tris	Fisher Scientific	BP152-1
Potassium hydroxide	Sigma	P5405
Sodium chloride	Fisher Scientific	S5886-10KG
n-octyl-β-D-glucopyranoside	Sigma	O8001
HEPES	Sigma	H4034
30% human serum albumin (HSA)	Sigma	12667-50ML-M
Streptavidin	Thermofisher	434,301
IgG1, Kappa from human myeloma plasma	Sigma	I5154-1MG
D-Biotin	Thermofisher	B20656
32% paraformaldehyde	VWR	100504-858
Normal donkey serum (NDS)	Jackson Immunoresearch Laboratories	017-000-121
Triton-X	Sigma	9036-19-5
190 Proof ethanol	Fisher Scientific	64-17-5
10 × PBS	Fisher Scientific	MT-46013CM

Continued

—cont'd

Reagent or resource	Source	Identifier
Storage glassware equipment		
Sample vials, Boroscilicate glass	VWR	14230-826
Experimental Models: Cell Lines		
Kappa CAR NK92MI cells	Generated in-house	
Equipment		
Lyophilizer	Labconco	Freezone 7740020
Sonicator	QSonica	Q500
Dialysis tubing	Fisher Scientific	08-670-5C
Vapro Osmometer	ELITechGroup	Model 5600
IBIDI Sticky-side VI 0.4 6-well chambers	IBIDI	80608
25x75mm glass coverslips	IBIDI	10812
Nikon A1R HD Confocal microscope	Nikon	
Software and Algorithms		
Image J (Fiji)	NIH	
Graphad Prism 9.3.1	Graphpad	
Other		
Machine learning-based CAR IS quality analysis	Developed in-house	See PLOS Computational Biology (Naghizadeh et al., 2022).

4 Materials and equipment
4.1 Chemicals, resources, and reagents

- 18:1(Δ-Cis) PC (DOPC) 1,2-dioleoyl-sn-glycero-3-phosphocholine (850375C, Avanti Polar Lipids)
- 18:1 Biotinyl Cap PE 1,2-dioleoylsn-glycero-3-phosphoethanolamine-N-(cap biotinyl) (870273C, Avanti Polar Lipids)
- Compressed Argon gas (AR UHP180LT230, Airgas)
- Glass vials for lipid storage (14230-826, VWR)
- Sulfuric acid (7664-93-9, ACROS organics)
- Hydrogen peroxide (386790, Sigma)
- Tris (BP152-1, Fisher Scientific)
- Sodium chloride (S5886-10KG, Fisher Scientific)
- 190 Proof ethanol (64-17-826, Fisher Scientific)
- Potassium hydroxide (P5405, Sigma)
- n-octyl-β-D-glucopyranoside (O8001, Sigma)

- Casein (C5890, Sigma-Aldrich)
- HEPES (H4034, Sigma-Aldrich)
- 10× PBS (MT-46013CM, Fisher Scientific)
- Human serum albumin (HSA) (12667-50ML-M, Sigma-Aldrich)
- Streptavidin (434301, ThermoFisher)
- Biotinylated-CD19-AF488 (AVI9269, R&D Systems)
- D-Biotin (B20656, Thermo Fisher)
- IgG1, Kappa from human myeloma plasma (B20656, Thermofisher)
- 32% paraformaldehyde (100504-858, VWR)
- Triton-X (9036-19-5, Sigma)
- Normal Donkey Serum (NDS) (017-000-121, Jackson Immunoresearch Laboratories)
- Alexa Fluor 405 Phalloidin—(A30104, Thermo Fisher)
- Fluorescently labeled secondary antibodies (Thermo Fisher)

4.2 Equipment and software
- Lyophilizer (Freezone 7740020, Labconco)
- Sonicator (Q500 sonicator, QSonica)
- Dialysis tubing (08-670-5C, Fisher Scientific)
- Vapro Osmometer (Model 5600, ELITechGroup)
- IBIDI Sticky-side VI 0.4, 6-well chambers (80608, Ibidi)
- 25x75mm glass coverslips (10812, Ibidi)
- Vacuum desiccator
- Nikon A1R
- Image J (FIJI)
- Graphpad Prism 9.3.1

Alternatives
- IBIDI Sticky-side VI 6-well chambers can be imaged using super-resolution STED microscopy or high-content imaging systems such as the InCell Analyzer from GE Healthcare or the ImageXpress® from Molecular Devices.
- His-tagged proteins can be used with 18:1 DGS-NTA(Ni) from Avanti Polar Lipids, Cat no. 790404P

5 Step-by-step method details
5.1 Liposome preparation
Timing: 4–24 h
1. Calculate desired amount of chloroform-suspended stock solutions of DOPC and Biotin-PE to make diluted stocks.
   a. Standard working concentrations are 400 µM DOPC and 80 µM Biotin-PE (2 mol%).
   b. For 10 mL of each lipid, 629 µL of DOPC (10 mg/mL) and 88 µL of Biotin-PE.

2. Break open lipid glass vials and aliquot desired amount into cleaned glass chromatography tubes.
3. Dry the chloroform by degassing with a stream of Argon for 5–10 min, depending on the volume. Lipids should form a thin, translucent layer. Cover with parafilm and store aliquots at −80 °C.
4. Place the degassed lipids in a lyophilizer to remove residual chloroform for at least 2 h to overnight.
5. To prepare 400 μM DOPC and 80 μM Biotin-PE, add 5 mL and 200 μL lipid suspension buffer into the DOPC and Biotin-PE glass tube, respectively.
6. Sonicate on ice for at least 15 min or until the solution becomes clear and no precipitate remains; 1 min on and 1 min off cycles.

5.2 Dialysis of liposomes
Timing: 3–4 days
1. Rehydrate 2 separate 6 mm dialysis tubing (10 mm flat diam/12–14,000 MW cut-off) by rinsing and soaking in water in a glass beaker for 2 min.
2. Microwave and boil the tubing for 5 min. Allow tubing to cool.
3. In a laminar flow hood, tie one end of tubing with a knot. Prior to adding the liposomes to the tubing, equilibrate with 1 mL lipid resuspension buffer and remove as much as possible.
4. Add liposomes to tubing and clamp the open ends by clamping 1 mm below the "water line" (see Note 1).
5. Immerse the samples in 1 L of lipid dialysis buffer. Add a stir-bar, degas the bottle with argon and seal with parafilm. Place in 4 °C overnight while changing the Tris buffer every 12–16 h.
6. After 4 buffer changes, cut dialysis tubing above the clip. Carefully open tubing and transfer lipids into a 2 mL screw-capped tube. Degas with argon purge and seal with parafilm. Store at 4 °C (see Notes 2–4).

Notes
1. Complete air exclusion in the dialysis tubing is crucial. This involves sacrificing small amounts of lipids when clamping the tubing.
2. Do not freeze or store lipid suspensions below 4 °C, as it will compromise the integrity of the lipid structure and fluidity.
3. Unilamellar liposomes can also be prepared using the alternative method with mini-extruder technique (Voss, Lee, Tian, Krzewski, & Coligan, 2018).
4. It is recommended to degas the lipid storage vials after use to preserve the lipid integrity longer.

5.3 Procedure for immunofluorescence of cells on lipid bilayers in flow chamber slides
Timing: 2–3 h
Experimental flow is illustrated in Fig. 1

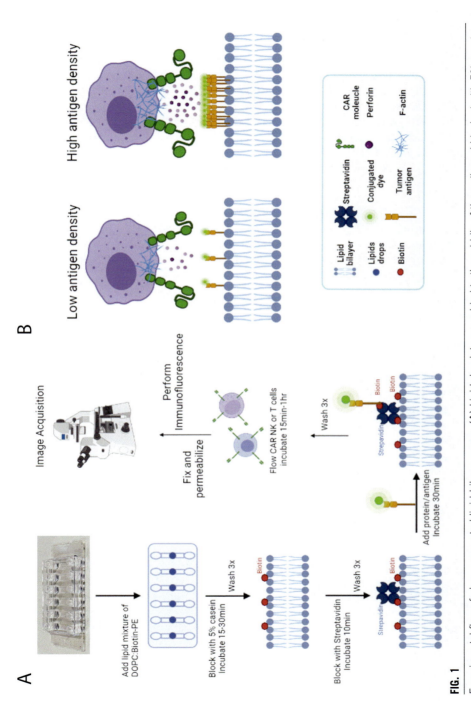

FIG. 1

Experimental flow of glass-supported lipid bilayer assay. (A) Lipid droplets are added to the middle of the wells and blocked with 5% casein. Biotin-PE is subjected to streptavidin flow-through followed by biotinylated proteins. After incubating with proteins, the lipid bilayer is blocked with D-biotin to saturate unbound streptavidin sites. CAR-NK or -T cells are added to the chamber and incubated at 37 °C. Cells are fixed, permeabilized, and stained for immunofluorescence microscopy. (B) A schematic for the effects of antigen density on immune synapse quality. Low antigen density is unable to induce efficient IS formation, whereas high antigen density has increased expression of IS quality parameters such as tumor antigen clustering, F-actin ring size, and perforin polarization.

1. Submerge glass coverslips in freshly made Piranha solution (3:1 sulfuric acid: hydrogen peroxide) for 30 min (see Note 1).
2. Rinse extensively with ultrapure RO (reverse osmosis) H_2O. Dry the slide with a stream of argon and wipe residual water at the corner with Kimwipes.
3. Mix 1:1 prepared DOPC and Biotin-PE. Add 2 μL of lipid mixture to the IBIDI slide and cover with cleaned glass coverslips (see Notes 2,3).
4. Flow 100 μL of 5% casein over lipids for 15–30 min at RT (see Note 4).
5. While incubating, place wash buffer in 50 mL tubes in a vacuum desiccator to remove gas bubbles (see Note 5).
6. Rinse with 1 mL wash buffer 3 times (see Note 6).
7. Flow 100 μL of 333 ng/mL streptavidin in PBS over lipids for 10 min at RT. Rinse with 3 mL wash buffer.
8. Add desired antigens with varying concentrations in 100 μL of PBS to determine the best functional range of the CAR activation. Incubate for 30 min at RT (see Note 7).
9. Flow 100 μL of 2.5 μM D-biotin in PBS over lipids for 10 min at RT. Rinse with 3 mL wash buffer (see Note 8).

Pause Point: Slides with embedded proteins on the lipid bilayer can be stored at 4 °C. Flow chambers will need to be supplemented with wash buffer every 1–2 days to prevent wells from drying out.

Timing: 20–24 h
10. While incubating proteins and D-biotin, prepare cells for lipid bilayer stimulation and immune synapse formation.
    a. Harvest and count NK92MI cells transduced with Kappa CAR (Fig. 2A).
    b. Wash 2–3 times with PBS and resuspend with $1.0–2.0 \times 10^6$ cells/mL in wash buffer.
11. Rinse lipid bilayer with 5 mL wash buffer to remove unbound proteins.
12. Flow 100 μL of Kappa CAR-NK92MI cells prepared in Step 10 over lipid bilayer and incubate 15–60 min at 37 °C[9].
13. Rinse 1 time with 1 mL wash buffer to remove unbound cells from the lipid bilayer.
14. Flow 500 μL 4% paraformaldehyde and incubate 15 min at RT. Rinse with 5 mL PBS[10].
15. Flow 500 μL permeabilization buffer and incubate for 30 min at RT (see Note 11).
16. While incubating with permeabilization buffer, dilute primary antibodies for immunological synapse quality synapse analysis (e.g., perforin, tumor antigen, and phospho-CD3ζ) to a 1:200 dilution in primary antibody dilution buffer. For F-actin staining, the phalloidin can be used with fixed samples.
17. Without washing, flow 100 μL of diluted antibody to the chamber and incubate 4 h to overnight at 4 °C.
18. After incubation, wash with 3 mL PBS.
19. Dilute secondary antibodies to 1:800–1:1000 in antibody dilution buffer and flow into the chamber for 2 h at RT.

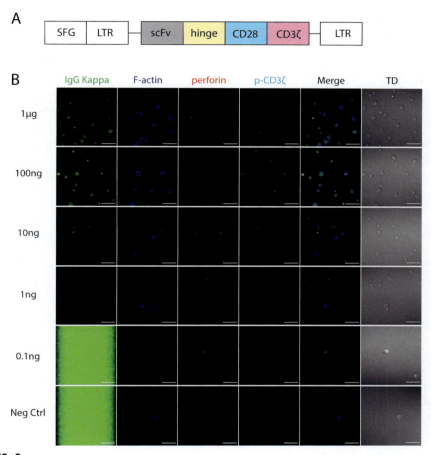

FIG. 2

Representative images of IgG1 Kappa CAR NK92MI immune synapse quality. (A) Schematic representation of recombinant retroviral vectors encoding second generation CAR. The construct contains the single-chain variable fragment (scFv) against IgG1 Kappa, CD28 transmembrane domain, and intracellular CD3ζ signaling domain. (B) Representative images of IgG1 Kappa CAR NK92MI immune synapse at 60× objective zoom. F-actin (Blue), IgG1 Kappa-AF488 (Green), perforin (Magenta), phospho-CD3ζ (Cyan), and transmitted image (TD). White bars indicate 50 μm.

20. Rinse with 5 mL PBS.
21. Image immediately or store at 4°C for 7–10 days without significant loss of signal.

Pause Point: The chamber will need to stay wet during storage. Replenish with sterile PBS every 2–3 days to avoid drying out the lipids and chamber. This prevents loss of signal and stabilizes lipid integrity.

Notes
1. A typical piranha solution mixture is 3 parts of concentrated sulfuric acid (93–98%) and 1 part of 30% hydrogen peroxide.
2. Two lipid droplets may be placed per well to increase the number of cells bound to the lipid bilayer.
3. Press down gently on the areas surrounding the flow chamber and draw a circle using a black permanent marker around lipid droplets to better find lipid bilayers during image acquisition.
4. Solution flow through the IBIDI chamber slide can be done using a pipette and an aspirator. Without introducing any bubbles, flow solution on one end of the slide with a pipette and aspirate on the opposite end simultaneously using an aspirator.
5. Removing bubbles is crucial to reducing damages to lipid bilayers as oxidation disrupts lipid integrity.
6. The standard washing volume is 3 mL. Use up to 5 mL for thorough washing of the wells.
7. Alternatively, protein densities on lipid bilayers can be calculated as described (Zheng, Bertolet, Chen, Huang, & Liu, 2015).
8. The slide can be stored 1–2 days in 4 °C. Ensure the chamber stays wet by adding 50 μL wash buffer.
9. Conditions such as time, antigen density, CAR constructs, and cell culture for the best immune synapse signal for a particular CAR must be optimized.
10. It is important to remove the paraformaldehyde as much as possible. Thus, we recommend 5 mL PBS per well.
11. Place permeabilization buffer in a desiccator to remove small bubbles for 5–10 min.

5.4 Immunofluorescence imaging using semi-automated image capture and analysis.

Timing: Acquisition 3–5 h, Analysis 4–8 h
1. Using the Nikon A1R with an automated stage mover and a 60× 1.4 NA objective, acquire images over the interface between the immune cell and the antigen. Starting with the highest density or concentration of antigen, set up fluorescent channels and its laser power.
2. To find the IS, find the highest intensity peak of the antigen channel and set it as current position. Next, set Z-stack as relative to the position.
3. Set Z-stacks to 0.25 μm per slice for 5 slices (1 μm total).
4. Pick 10–20 points within the lipid bilayer identified by the black marks made from step 3.3.3.
5. Start acquiring via automated stage to capture the immune synapse interface.
6. Use Image analysis software provided by Nikon, ImageJ, or our machine learning (ML)-based image segmentation and quantification (Naghizadeh et al., 2022).
7. Count the mean fluorescence intensity (MFI) of each channel of each cell.
8. Plot MFI and determine immune synapse quality based on CAR constructs.

5.5 Machine learning-based CAR IS quality analysis
Timing: Analysis less than 30 min
1. Export microscopic images to 16-bit and 8-bit formats.
2. The images processing can be done using our ML-based quantification software (Naghizadeh et al., 2022). This requires Python programming language and its prominent libraries such as PyTorch and OpenCV.
3. Quantification involves the detection of CAR-NK IS surface boundaries, calculating total fluorescence intensity and MFI. The MFI is obtained by dividing the total intensity of a CAR-NK IS over the area of its surface boundary.
4. To detect the surface boundaries, ML software requires the 8-bit version of the images to improve the software detection of the IS area and interface.
5. Use the tumor antigen channel to select the z-stack with the best/highest intensities.
6. F-actin channel is used for surface boundaries which leads to instance segmentation masks and cell boundaries.
7. The boundaries obtained from F-actin channels are applied to other fluorescent channels to quantify intensities within the instance segmentation masks.
8. To obtain the unmodified version of the pixel values, 16-bit images are used to evaluate the intensity of the CAR-NK IS images. Pixels from the 8-bit images are correlated with 16-bit images, and the intensity values from 16-bit images are used for calculations.

6 Expected outcomes

Image acquisition of the chamber slide should expect a decreasing intensity of the IS parameters with less antigen on the lipid bilayer (Fig. 2B). Negative controls (such as the data from outside of the lipid bilayer area) and any antigen amount unable to elicit CAR recognition and antigen clustering will show a bright fluorescent image due to auto-contrast increasing the background noise, indicating the lack of any signals in the channel. Colors are added to the original grayscale images for better representation and to make the cells visible to human eyes. Color adjustments do not affect the real analysis. In our case, negative control did not form any IS and 0.1 ng of antigen was too low for antigen clustering, as indicated by the bright green fluorescent channel after image processing.

Quantification of the immune synapse quality parameters manually or automatically should expect to see an antigen-dependent trend with the highest antigen amount having higher MFI (better IS quality) and more cells bound to lipid bilayers compared to low antigen amounts and negative controls (Fig. 3). As expected, the highest concentration of IgG Kappa (1 μg) showed the highest number of cells bound (Fig. 3A), area (Fig. 3B), F-actin MFI and tumor antigen MFI (Fig. 3C). Interestingly, the MFI of perforin and phospho-CD3ζ were higher in the 10 ng compared to others. This could be due to an underlying biological or technical issue with the assay. Thus, repeats are extremely important when running the protocol

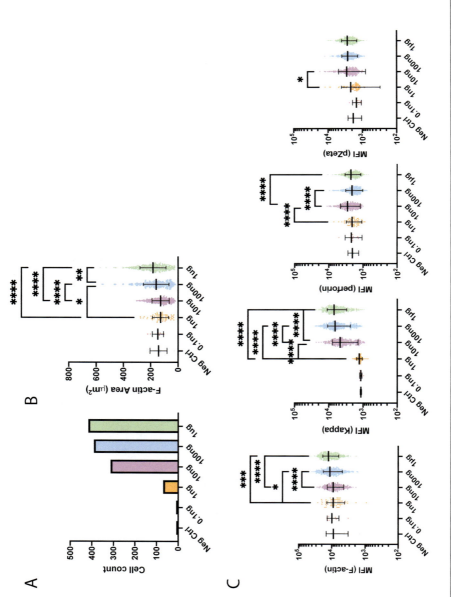

FIG. 3

Quantification of the CAR IS quality on CAR-modified cells stimulated on the glass-supported planar lipid bilayer. IS quality assessment of IgG1 Kappa-CAR-NK92MI cells with varying antigen density on the glass-supported planar lipid bilayer. (A) Number of cells bound to lipid bilayer after wash and fix. (B) The area of the synapse defined by the covered area of F-actin ring. (C) Quantification of the IS parameters including intensity of F-actin, IgG1 Kappa, perforin, and phospho-CD3ζ (pCD3ζ). One-way ANOVA with post-hoc Tukey test was performed, and statistical significance was graphed using *$P < 0.05$, **$P < 0.01$, ***$P < 0.001$, ****$P < 0.0001$. Experiment was performed once with cell numbers ranging from 8 to 416.

described in this paper. Lastly, the accuracy of this method deteriorates when the cell numbers are too low. For example, negative control, 0.1 ng, and 1 ng only have 8, 9, and 69 cells that were bound, respectively. The spread of the intensities becomes larger, and the confidence interval increases. Thus, it is typically necessary to have more 200 cells per treatment for accurate detection and quantification of the immune synapse parameters.

Overall, the standardized protocol presented in this manuscript aims to help find an optimal range of working tumor antigen amounts and conditions to characterize the interaction between CAR-modified cells and the antigen. Based on our analysis of the IS parameters with IgG_1 Kappa protein and Kappa CAR NK92MI cells, the results suggest that a range of 1 ng to 100 ng of antigen would be a good starting point to characterize or optimize CAR constructs.

7 Statistical analysis

One-way ANOVA with post-hoc Tukey test was performed, and statistical significance was graphed by $*P<0.05$, $**P<0.01$, $***P<0.001$, $****P<0.0001$. The experiment was performed once with cell numbers ranging from 8 to 416.

8 Limitations

While the glass-supported planar lipid bilayers have advantages in that they can mimic the fluidity of the cell membrane and allow for direct visualization of the IS through high-resolution imaging, the use of the assay in a high-throughput format has been limited. Our glass-supported lipid bilayer assay is currently limited to a 6-well chamber slide and manual identification and imaging of the IS. Another major limitation of the assay is the variability between replicates, personnel, and experimental conditions. At least three experiments with three separate slides are needed to confirm true CAR IS quality phenotypes with well-designed positive and negative controls within the same assay. Thus, the low-throughput nature of the assay is time-inefficient which limits the number of samples for comparison per experiment.

Michael Dustin and colleagues have developed an automated high-throughput method with a modified version of the lipid bilayer assay (Valvo et al., 2017). However, the protocol requires sophisticated machinery, including liquid handlers and high-content imaging equipment such as the GE InCell Analyzer. The number of parameters is also limited in that we can only evaluate 4–5 proteins of interest per sample. Optimization and fluorescence bleed-throughs become technically challenging with more parameters that limit the number of parameters per sample to 4–5 biomarkers of interest. Furthermore, lipid bilayer assays lack target cells which do not allow for full investigation into signaling pathways and interactions of a variety of co-receptors present in the IS. Alternatively, our lab has developed a vertical-cell pairing (VCP) system that enables IS imaging in a horizontal focal plane with both fixed- and live-cell imaging (Jang et al., 2015). Importantly, this device allows for interrogation of the IS structure from more than 8000 effector and target cell pairs per experimental run. More research and development on biodevices that reveal the interaction between effector and target cells will be crucial in the future to use IS quality as a biomarker in the clinics.

References

Dustin, M. L., & Long, E. O. (2010). Cytotoxic immunological synapses. *Immunological Reviews*, *235*(1), 24–34. https://doi.org/10.1111/j.0105-2896.2010.00904.x.

Dustin, M. L., Starr, T., Varma, R., & Thomas, V. K. (2007). Supported planar bilayers for study of the immunological synapse. *Current Protocols in Immunology*, *76*, 18.13.1–18.13.35. https://doi.org/10.1002/0471142735.im1813s76 (Chapter 18, Unit 18 13).

Graf, B., Bushnell, T., & Miller, J. (2007). LFA-1-mediated T cell costimulation through increased localization of TCR/class II complexes to the central supramolecular activation cluster and exclusion of CD45 from the immunological synapse. *Journal of Immunology*, *179*(3), 1616–1624. https://doi.org/10.4049/jimmunol.179.3.1616.

Grakoui, A., Bromley, S. K., Sumen, C., Davis, M. M., Shaw, A. S., Allen, P. M., et al. (1999). The immunological synapse: A molecular machine controlling T cell activation. *Science*, *285*(5425), 221–227. https://doi.org/10.1126/science.285.5425.221.

Halim, L., Das, K. K., Larcombe-Young, D., Ajina, A., Candelli, A., Benjamin, R., et al. (2022). Engineering of an avidity-optimized CD19-specific parallel chimeric antigen receptor that delivers dual CD28 and 4-1BB co-stimulation. *Frontiers in Immunology*, *13*, 836549. https://doi.org/10.3389/fimmu.2022.836549.

Jang, J. H., Huang, Y., Zheng, P., Jo, M. C., Bertolet, G., Zhu, M. X., et al. (2015). Imaging of cell-cell communication in a vertical orientation reveals high-resolution structure of immunological synapse and novel PD-1 dynamics. *Journal of Immunology*, *195*(3), 1320–1330. https://doi.org/10.4049/jimmunol.1403143.

Liu, D., Badeti, S., Dotti, G., Jiang, J. G., Wang, H., Dermody, J., et al. (2020). The role of immunological synapse in predicting the efficacy of chimeric antigen receptor (CAR) immunotherapy. *Cell Communication and Signaling: CCS*, *18*(1), 134. https://doi.org/10.1186/s12964-020-00617-7.

Liu, D., Peterson, M. E., & Long, E. O. (2012). The adaptor protein Crk controls activation and inhibition of natural killer cells. *Immunity*, *36*(4), 600–611. https://doi.org/10.1016/j.immuni.2012.03.007.

Majzner, R. G., Rietberg, S. P., Sotillo, E., Dong, R., Vachharajani, V. T., Labanieh, L., et al. (2020). Tuning the antigen density requirement for CAR T-cell activity. *Cancer Discovery*, *10*(5), 702–723. https://doi.org/10.1158/2159-8290.CD-19-0945.

Markey, K. A., Gartlan, K. H., Kuns, R. D., MacDonald, K. P., & Hill, G. R. (2015). Imaging the immunological synapse between dendritic cells and T cells. *Journal of Immunological Methods*, *423*, 40–44. https://doi.org/10.1016/j.jim.2015.04.029.

Monks, C. R., Freiberg, B. A., Kupfer, H., Sciaky, N., & Kupfer, A. (1998). Three-dimensional segregation of supramolecular activation clusters in T cells. *Nature*, *395*(6697), 82–86. https://doi.org/10.1038/25764.

Naghizadeh, A., Tsao, W. C., Hyun Cho, J., Xu, H., Mohamed, M., Li, D., et al. (2022). In vitro machine learning-based CAR T immunological synapse quality measurements correlate with patient clinical outcomes. *PLoS Computational Biology*, *18*(3), e1009883. https://doi.org/10.1371/journal.pcbi.1009883.

Obino, D., Diaz, J., Saez, J. J., Ibanez-Vega, J., Saez, P. J., Alamo, M., et al. (2017). Vamp-7-dependent secretion at the immune synapse regulates antigen extraction and presentation in B-lymphocytes. *Molecular Biology of the Cell*, *28*(7), 890–897. https://doi.org/10.1091/mbc.E16-10-0722.

Purbhoo, M. A., Liu, H., Oddos, S., Owen, D. M., Neil, M. A., Pageon, S. V., et al. (2010). Dynamics of subsynaptic vesicles and surface microclusters at the immunological synapse. *Science Signaling*, *3*(121), ra36. https://doi.org/10.1126/scisignal.2000645.

Singh, N., Frey, N. V., Engels, B., Barrett, D. M., Shestova, O., Ravikumar, P., et al. (2021). Antigen-independent activation enhances the efficacy of 4-1BB-costimulated CD22 CAR T cells. *Nature Medicine*, *27*(5), 842–850. https://doi.org/10.1038/s41591-021-01326-5.

Valvo, S., Mayya, V., Seraia, E., Afrose, J., Novak-Kotzer, H., Ebner, D., et al. (2017). Comprehensive analysis of immunological synapse phenotypes using supported lipid bilayers. *Methods in Molecular Biology*, *1584*, 423–441. https://doi.org/10.1007/978-1-4939-6881-7_26.

Voss, O. H., Lee, H. N., Tian, L., Krzewski, K., & Coligan, J. E. (2018). Liposome preparation for the analysis of lipid-receptor interaction and Efferocytosis. *Current Protocols in Immunology*, *120*, 14.44.1–14.44.21. https://doi.org/10.1002/cpim.43.

Xiong, W., Chen, Y., Kang, X., Chen, Z., Zheng, P., Hsu, Y. H., et al. (2018). Immunological synapse predicts effectiveness of chimeric antigen receptor cells. *Molecular Therapy*, *26*(4), 963–975. https://doi.org/10.1016/j.ymthe.2018.01.020.

Zheng, P., Bertolet, G., Chen, Y., Huang, S., & Liu, D. (2015). Super-resolution imaging of the natural killer cell immunological synapse on a glass-supported planar lipid bilayer. *Journal of Visualized Experiments*, *96*. https://doi.org/10.3791/52502.

CHAPTER 12

Potency monitoring of CAR T cells

Dongrui Wang[a,†], Xin Yang[a], Agata Xella[a], Lawrence A. Stern[b], and Christine E. Brown[a,*]

[a]T Cell Therapeutics Research Laboratories, Cellular Immunotherapy Center, Department of Hematology and Hematopoietic Cell Transplantation, City of Hope, Duarte, CA, United States
[b]Department of Chemical, Biological and Materials Engineering, University of South Florida, Tampa, FL, United States
*Correspondence author: e-mail address: cbrown@coh.org

Chapter outline

1	Introduction	174
2	Preparation of cells	175
3	Setup co-culture for extended long-term killing (ELTK) assay (Assay #1)	176
4	Setup co-culture for *re*-challenge assay (Assay #2)	178
5	General procedures of flow cytometry analysis	179
6	Analysis of CAR T cell killing and T cell counts by flow cytometry	179
7	Analysis of CAR T cell phenotypes by flow cytometry	181
8	Quantification of cytokine production by CAR T cells	185
9	Concluding remarks	185
10	Notes	186
	References	187

Abstract

The effector potency of chimeric antigen receptor (CAR) T cell therapeutic products is essential to their clinical antitumor responses, and potency monitoring is a critical quality control method for CAR T cell therapy platforms. While many *in vitro* assays enable high-throughput assessment of CAR T cell cytotoxicity, it has been challenging for these assays to reflect the *in vivo* therapeutic effect due to their nature as short-term methods that fail to recapitulate the high tumor burden environment. Here, we describe two *in vitro* co-culture methods to evaluate

---

[†]Current address: Department of Immunology, University of Texas, MD Anderson Cancer Center, Houston, TX, United States

CAR T cell recursive killing potential at high tumor cell loads. In these assays, long-term cytotoxic function and proliferative capacity of CAR T cells are examined *in vitro* over 7 days. Further, these assays are coupled with profiling CAR T cell expansion, cytokine production and phenotypes. These methods provide a facile approach to assess CAR T cell potency and to elucidate the functional variations across different CAR T cell products.

1 Introduction

Chimeric antigen receptor (CAR)-engineered T cells have been used to treat various types of tumors, with particularly promising clinical outcomes against B cell malignancies (Brentjens et al., 2013; Grupp et al., 2013; Lee et al., 2015; Porter et al., 2015). While the efficacy of targeting other tumors by CAR T cells still shows mixed results, there continues to be rigorous optimization of CAR T cell products and combination therapies (Akhavan et al., 2019; Fesnak, June, & Levine, 2016; Priceman, Forman, & Brown, 2015). Of note, clinical research on CAR T cell-treated patients has revealed a correlation between the intrinsic properties/fitness of the CAR T cell product and clinical response rates (Deng et al., 2020; Fraietta, Lacey et al., 2018), illustrating the importance of delineating bespoke CAR T cell products and further augmenting functional potency of the therapeutic cells. Thus, it is critical to include a robust potency-monitoring strategy as part of the characterization of novel CAR designs and patient-specific cell products.

It has been challenging, however, to assess the therapeutic potential of CAR T cells to inform clinical translation, especially in a high-throughput manner. The current gold-standard for assessing functional potency is to compare titrated doses of CAR T cell antitumor potency in immunodeficient mice bearing human tumor xenografts as a "stress-test" (Brown et al., 2017; Cherkassky et al., 2016; Eyquem et al., 2017; Long et al., 2015; Priceman et al., 2018), which are extremely labor-intensive, time-consuming, and expensive. This approach can also be constrained by the accessibility of mouse strains, animal care facilities and animal-handling techniques. Therefore, there is a need to develop more convenient *in vitro* assays allowing for efficient readouts of effector activity which faithfully reflect the *in vivo* antitumor function of these T cells.

Previously, *in vitro* cytotoxic evaluation of T cells has focused on the detection of degranulation, short-term cytokine production and short-term lysis of target cells (*i.e.*, chromium release assay (Brunner, Mauel, Cerottini, & Chapuis, 1968), luciferase killing assay (Vishwanath et al., 2005)). These assays usually provide readout within 24 h and are informative for defining CAR T cell specificity and redirected target recognition, however, they often fail to reflect *in vivo* antitumor potential of engineered T cells (Cherkassky et al., 2016; Gattinoni et al., 2005; Long et al., 2015). Such inconsistency is likely the result of short-term and high effector: target (E:T) ratios used in these *in vitro* assays, which are, thus, unable to reflect T cell exhaustion, an important barrier for effective CAR T cell treatment (Malandro et al., 2016). Meanwhile, most short-term *in vitro* killing assays also do not characterize differences in T cell proliferation and phenotypic alterations post-stimulation, two important parameters correlated with clinical responses (Lee et al., 2015). Thus, the appropriate *in vitro* assay would need to recapitulate conditions of high tumor burden, induction of T cell exhaustion and provide readouts other than direct killing of target cells.

Here we describe two *in vitro* strategies to monitor the potency of therapeutic CAR T cells: the extended long-term killing assay (ELTK) and the tumor rechallenge assay. With the ELTK, CAR T cell potency can be evaluated by exposing effector cells to excess amounts of tumor cells, mimicking the *in vivo* condition of high tumor burden. Further, the tumor rechallenge assay is able to analyze the dynamics of CAR T cell cytotoxicity and other phenotypic and functional readouts, as different T cell effector activity parameters (*e.g.*, killing potency, expansion, cytokine production and surface markers) can be simultaneously examined (Wang et al., 2019). The results generated from these assays correlate well with *in vivo* antitumor potency (Wang et al., 2018) and can be readily adapted to other CAR T cell platforms. This is especially important for quality assessment of clinical CAR T cell products.

2 Preparation of cells

a. Preparation of co-culture media (Note 1): DMEM:F12 (MediaTech, Manassas, VA) containing 10% fetal calf serum (FCS) (Hyclone, Logan, UT) (Note 2); preparation of T cell media: X-VIVO 15 (Lonza, Basel, Switzerland) containing 10% FCS, 70 IU/mL rhIL-2 (Novartis Oncology, East Hanover, NJ) and 0.5 ng/mL rhIL-15 (CellGenix, Freiburg im Breisgau, Germany) (Note 3).
b. Preparation of effector T cells
   i. Thaw CAR T cells by transferring one frozen vial (kept on ice) into a 37 °C water bath. Keep the vial moving in the bath to ensure an even thaw until a small amount of ice remains (1–3 min). Add 1 mL pre-warmed T cell media and mix gently using a transfer pipette.
   ii. Wash once with 10 mL of pre-warmed T cell media, resuspend at 6×10^5 cells/mL in T cell media and culture overnight (Note 4)
   iii. Quantify the percentage of CAR positive cells in the product by either performing flow cytometry on the thawed T cells or, alternatively, obtain T cell manufacturing information regarding % CAR expression (Note 5).
   iv. Determine T cell concentration by a cell viability counter and resuspend CAR T cells in co-culture media at the specified concentration (cells/mL) (Sections 3.b.iv and 4.b.iv).
c. Preparation of target cells (Note 6)
   i. Harvest and make a single-cell suspension of tumor cells by the appropriate dissociation agent as recommended by the cell line producer.
   ii. Pre-warm co-culture media in a 37 °C water bath.
   iii. Determine cell concentration by a cell viability counter and resuspend target cells in co-culture media at the specified concentration (cells/mL) (Note 7).

3 Setup co-culture for extended long-term killing (ELTK) assay (Assay #1)

a. Optimal E:T ratio must be empirically determined based on T cell function and tumor growth kinetics (Note 8). Here we detail our standard protocol for comparing the potency of patient CAR T cell products performed at a 1:50 E:T ratio over 7-days without exogenous cytokine addition (Fig. 1).

b. Day 0 (Day of plating):
  i. Determine the plate map for a 96-well flat bottom plate according to the CAR T cell samples to be analyzed and include a minimum of three replicate wells per experimental condition for each time point.
  ii. Harvest and resuspend single cell suspension of tumor cells according to Sections 2.c.i-iii to reach a final concentration of 2.5×10^5 cells/mL of single cell suspension in co-culture media (Note 7).
  iii. Pipette 100 µL of diluted tumor cells into each well of a 96-well flat-bottom tissue culture plate for a total of 2.5×10^3 tumor cells per well.
  iv. Harvest and resuspend T cells according to Sections 2.b.i-iii at a final concentration of 5×10^3 CAR-expressing cells/mL in co-culture media (CAR+ cells/mL = 5×10^3 cells/mL ÷ % CAR-positivity) (Note 7).
  v. Pipette 100 µL of diluted CAR T cells into each well that contains tumor cells and mix well for a total of 500 CAR T cells per well.

c. Day 3 or 4 (Media change):
  i. Carefully discard (or collect) 100 µL of supernatant from the top of each well without disturbing the cells (Notes 9). The supernatant can be transferred to a new round-bottom 96-well plate and kept at −20 °C for future analysis.
  ii. Without disturbing the pellet, gently supplement with 100 µL warm, fresh co-culture media.

d. Day 7 (Cell harvesting):
  i. (Optional) Take representative images of each well using a brightfield microscope.
  ii. Pre-warm 0.05% trypsin-EDTA solution (Corning, Corning, NY) in a 37 °C water bath.
  iii. Gently remove all media from wells using a pipette and transfer into a new round-bottom 96-well plate. Pipette gently to resuspend free-floating CAR T cells and avoid disturbing attached tumor cells for best results.
  iv. Pipette 50 µL trypsin-EDTA into the wells of the culture plate to digest remaining tumor cells and incubate at 37 °C for about 5 min.
  v. Under the microscope, confirm all cells have detached from the bottom of the plate.
  vi. Pipette around the well bottom to resuspend detached cells, transfer trypsin-EDTA containing detached cells to the corresponding wells of the

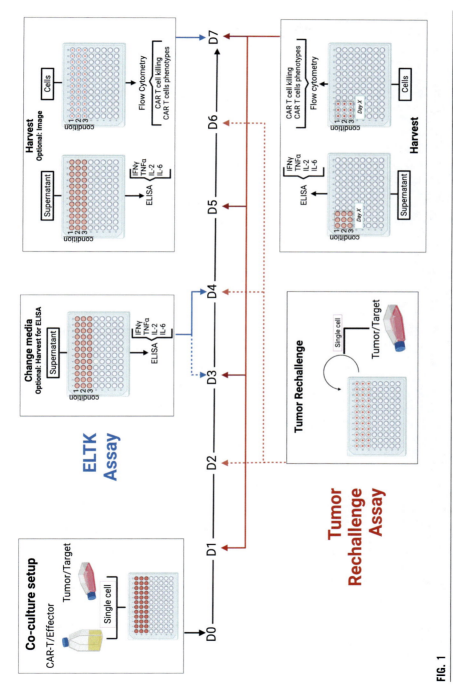

FIG. 1

Assay Schema. Overview of ELTK (Section 3) and Tumor Rechallenge (Section 4) assays showing experimental set-up for CAR T cell effector function evaluation using high-tumor burden co-culture assays. Image created with BioRender.com

round-bottom 96-well plate containing free-floating T cells. Alternatively, cell populations could be transferred to separate plates for independent analysis.

4 Setup co-culture for *re*-challenge assay (Assay #2)

a. The optimal effector to target (E:T) ratio must be empirically determined based on CAR T cell characteristics and tumor growth kinetics (Note 8). Here, we detail our standard protocol for comparing the potency of patient-derived CAR T cell products against tumor lines performed at a starting E:T ratio of 1:4, with 3 subsequent tumor additions at a 1:8 E:T ratio based on the starting CAR T cell input number (Fig. 1).

b. Day 0 (Day of plating):
   i. Determine the 96-well flat bottom plate map according to the CAR T cell samples to be analyzed and include a minimum of three replicate wells per experimental condition for each time point (Note 10).
   ii. Harvest target cells following Sections 2.c.i-iii, and resuspend at 1.6×10^5 cells/mL into co-culture media (Note 7).
   iii. Pipette 100 μL of diluted tumor cells into each well of a 96-well flat-bottom tissue culture plate for a total of 1.6×10^3 cells per well.
   iv. Harvest T cells following Sections 2.b.i-iii, and resuspend at 4×10^4 CAR-expressing cells/mL into co-culture media (Note 7).
   v. Pipette 100 μL of diluted CAR T cells into each well that contains tumor cells and mix well for a total of 4×10^3 cells per well

c. Days 2, 4 and 6 (Tumor re-challenge):
   i. Harvest target tumor cells following Sections 2.c.i-iii, resuspend 3.2×10^5 cells/mL into co-culture media.
   ii. Use a pipette to resuspend the cells in all the wells of the assay plate and transfer the cells into a new, round-bottom 96-well plate.
   iii. Plate 100 μL of fresh tumor cells into each well of the assay plate from where the cells have been transferred out.
   iv. Centrifuge the round-bottom 96-well plate at $300 \times g$ at 4 °C for 4 min, discard supernatant (Note 9).
   v. Resuspend the cell pellets in 100 μL of co-culture media and transfer them back into the corresponding wells of the assay plate containing fresh tumor cells.

d. Days 1, 3, 5 and 7 (Harvest for analysis):
   i. Pre-warm 0.05% trypsin-EDTA solution in a 37 °C water bath
   ii. From the assay plate, determine the wells that need harvesting. Transfer all the media and cells from these wells into a new, round-bottom 96-well plate (analysis plate). Do not rinse the wells in the assay plate at this point (Notes 9–10).

    iii. Centrifuge the analysis plate at 300 × g at 4 °C for 4 min. Carefully transfer 100–120 μL of supernatant from the top of each well into a new, round-bottom 96-well plate (supernatant plate) being careful to not disturb the pelleted cells (if the supernatant is not needed, media can be removed by a quick flick of the plate (Note 11)). Label the plate appropriately and keep the harvested supernatant at −20 °C for further analysis (see Section 8).
    iv. Pipette 50 μL trypsin-EDTA into the harvested wells of the assay plate to detach remaining cells and incubate at 37 °C for 5 min (Note 9).
    v. Using a light microscope, confirm the cells have detached from the bottom of the assay plate.
    vi. Pipette around the well bottom to resuspend detached cells. Transfer trypsin-EDTA containing detached cells to the corresponding wells of the analysis plate (Note 10).
    vii. Place the assay plate back in the 37 °C incubator for the next timepoint(s).

5 General procedures of flow cytometry analysis

a. Prepare FACS staining solution (FSS): HBSS (Irvine Scientific, Santa Ana, CA), 2% FCS, NaN_3 (0.5 g/500 mL) (Sigma-Aldrich, St Louis, MO).
b. Centrifuge the round-bottom 96-well plates from Sections 3.d.vi and 4.d.vi at 300 × g at 4 °C for 4 min, discard supernatant (Note 12). There is no need to keep the culture sterile at this point.
c. Add 200 μL/well of FSS to wash the cells, centrifuge at 300 × g at 4 °C for 4 min, discard supernatant.
d. Resuspend cells in 100 μL/well FSS containing antibodies (see Sections 6 and 7), stain cells at 4 °C for 30 min.
e. Add 100 μL/well FSS to cells, centrifuge at 300 × g at 4 °C for 4 min, discard supernatant.
f. Add 200 μL/well of FSS to wash the cells, centrifuge at 300 × g at 4 °C for 4 min, discard supernatant.
g. Resuspend cells with 100–200 μL/well FSS with 500 ng/mL DAPI, analyze samples using a flow cytometer. At this step, it is crucial that the resuspension volumes are precise, as the cell/mL counts will be used to calculate killing and proliferation (Section 6, Figs. 3–4).

6 Analysis of CAR T cell killing and T cell counts by flow cytometry

a. Following, we describe general antibody panel and method to evaluate CAR T cell performance in both assays in terms of killing and proliferation. This panel should be adapted to specific experimental conditions.

180 CHAPTER 12 Potency monitoring of CAR T cells

b. Antibodies used for flow cytometric analyses (in 100 μL total volume): anti-human-CD45-PerCP (3 μL/sample), anti-human-CD19-PE Cy7 (1 μL/sample) (BD Biosciences, San Jose, CA).
c. Target cells and CAR T cells can be distinguished by anti-CD45 staining (Fig. 2).

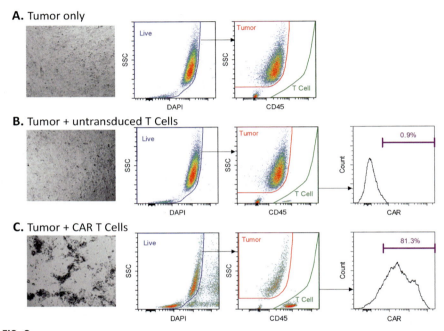

FIG. 2

Example of flow cytometry gating strategy. The human patient-derived glioblastoma tumor line, PBT030, was co-cultured with IL13Rα2-targeted CAR T cells in an ELTK assay, and tumor and T cells were assessed by flow cytometry at assay end-point (day 7). Left images show wells prior to harvest. Dot plots show representative gating of live cells (DAPI-negative) and sub-gating of residual tumors (CD45-negative) and T cells (CD45-positive) from 3 conditions. (A) Gating strategy for tumor only condition. Tumor cells gated first on live gate based on DAPI-negative, followed by CD45-negative side scatter high. (B) Gating strategy for tumor and untransduced T cells. T cells gated first on live gate based on DAPI-negative, followed by CD45-positivity (C) Gating strategy for tumor and CAR-T cell co-culture. CAR-positive CD3 cells were detected with anti-CD19, a transduction marker included within the CAR cassette. CAR-positivity gating was determined using the untransduced T cells (B) as the negative control.

d. CAR T cells can be gated by anti-CAR staining using the CAR target or staining against a reporter transgene that is part of the CAR construct (Note 12). Figure 3 shows the CAR gating strategy by staining a truncated CD19 transgene, which is the reporter transgene for the CAR T cells. **CAR T cell expansion** will then be calculated as fold expansion compared with the input cell numbers (Fig. 3–4).
e. **CAR T cell killing** can be evaluated by quantifying viable tumor cells. For Assay #1, percentage of killing can be calculated based on tumor cells undergoing identical culture time without CAR T cells (Fig. 3). For Assay #2, it is recommended that remaining viable tumor cell counts be plotted to reflect CAR-mediated killing (Fig. 4).
f. Proceed with T cell evaluation as suggested in Section 7.

7 Analysis of CAR T cell phenotypes by flow cytometry

a. Herein, we outline a general antibody panel to evaluate CAR T cell memory, activation and exhaustion marker expression. Changes to this panel would be expected to be project specific.
b. Antibodies used for flow cytometric analyses (in 100 μL total volume): anti-human-CD45-PerCP, anti-human-CD19-PE Cy7 (see Section 6), anti-human-CD8 APC Cy7 (2 μL/sample), anti-human-CD137 PE (5 μL/sample), anti-human-CD69 APC (3 μL/sample), anti-human-CD62L APC (5 μL/sample), anti-human-CD45RA PE (3 μL/sample) (BD Biosciences, San Jose, CA), anti-human-LAG-3 PE (1 μL/sample), anti-human-TIM-3 APC (1 μL/sample) (eBioscience, San Diego, CA) and anti-human-PD-1 APC Cy7 (1 μL/sample) (BioLegend, San Diego, CA).
c. Analyses should be performed on pre-gated CAR T cells, appropriate control stains including a fluorescence minus one (FMO) to ensure true gating. T cell activation can be assessed by co-expression of CD137 and CD69; T cell exhaustion can be assessed by co-expression of LAG-3, TIM-3 and PD-1 (Note 13). T cell memory subsets can be evaluated by CD62L and CD45RO: naïve/stem cell memory (CD62L+,CD45RO-), central memory (CD62L+, CD45RO+) and effector memory (CD62L-, CD45RO+) (Note 14; Fig. 5).

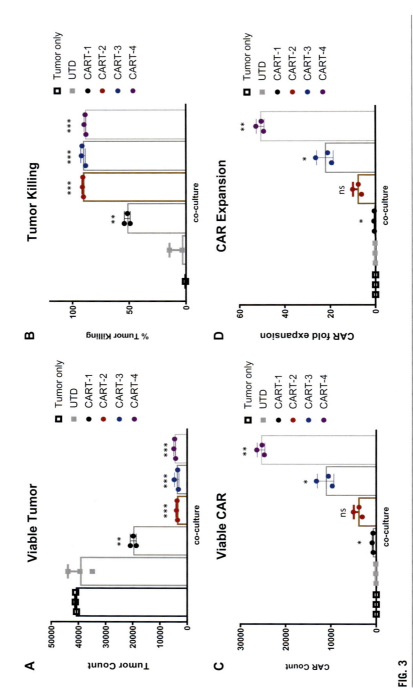

FIG. 3

ELTK Assay. IL13Rα2-CAR T cell products (CART) were co-cultured with PBT030 at an E:T ratio of 1:50 and residual tumors and viable CAR T cells were quantified at day 7. (A) Viable tumor was enumerated by flow cytometry at day 7 as $\frac{\text{tumor cells per well}}{\text{volume analyzed}} \times \text{final resuspension volume}$. (B) Tumor killing is calculated as $K = \left(1 - \frac{\text{count of well } K}{\text{average count of tumor only}}\right) \times 100\%$. (C) Viable CART is represented by absolute CAR count from each condition at assay end point. (D) CART expansion is calculated as dividing residual CART number by input CART number. (mean ± SD; n=3 replicated assessments; one-way ANOVA, $*P<0.033$, $**P<0.02$ and $***P<0.01$ as compared to untransduced (UTD) T cell group).

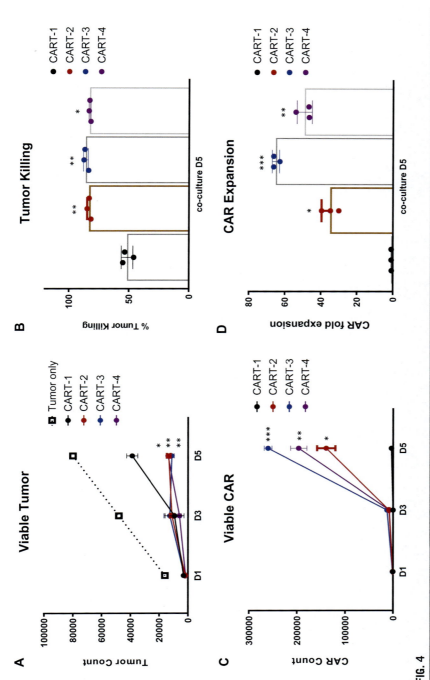

FIG. 4

Tumor Rechallenge Assay. IL13Rα2-CAR T cell products (CART) were co-cultured with PBT030 at an initial E:T ratio of 1:4, followed by tumor rechallenge at an E:T of 1:8 on days 2 and 4 (based on original input CAR T cells). (A) Viable tumor was enumerated by flow cytometry on day 1, 3 and 5. Dotted tumor only line represents theoretical value based on input tumor cells. (B) Tumor killing is calculated as $K = \left(1 - \frac{count\ of\ well\ K}{assay\ tumor\ input}\right) \times 100\%$. (C) Viable CAR is represented by absolute CAR count from each condition at assay end point. (D) CAR expansion is calculated as dividing residual CAR count by input CAR number. (mean ± SD; n = 3 replicated assessments; one-way ANOVA, *P < 0.033, **P < 0.02 and ***P < 0.01 as compared to CART-1 group).

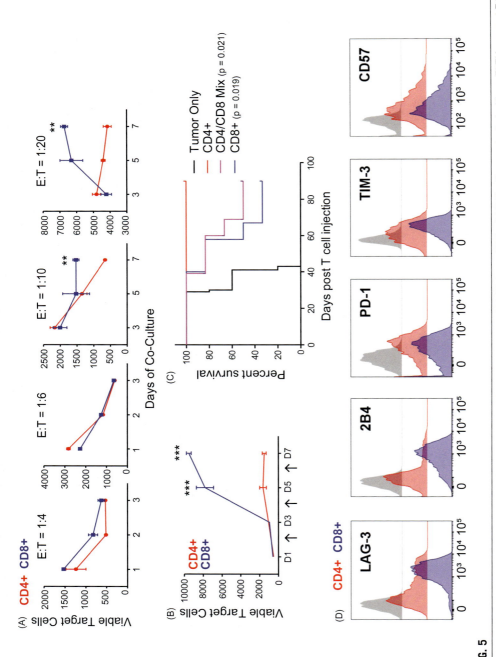

FIG. 5

ELTK and Rechallenge assays are predictive of *in vivo* antitumor responses. (A) CD4+ or CD8+ CAR T cells were tested with an ELTK assay at E:T ratio of 1:20, and the numbers of viable tumor cells were enumerated at the denoted time points. (mean ± SD; n = 3 replicates per time point. **$P < 0.01$ using an unpaired Student's t-test). (B) CD4+ or CD8+ CAR T cells were tested with a rechallenge assay and remaining viable tumor cell numbers were quantified at the indicated time points (arrows indicate tumor rechallenge). (mean ± SD; n = 3 replicates per data point. ***$P < 0.001$, unpaired Student's t-test comparing the CD4+ and CD8+ groups at the indicated time points). (C) Representative histograms of inhibitory receptor staining on CD4+ and CD8+ CAR T cells after rechallenge. (D) Survival of orthotopic tumor-bearing mice treated with different groups of CAR T cells. Kaplan-Meier survival analysis with p-values when compared with the CD4+ CAR T cell group using the Log-rank (Mantel Cox) test.

Figure adapted from original data published in Wang et al., 2018.

8 Quantification of cytokine production by CAR T cells

a. Thaw harvested supernatant from Section 4.d.iii at room temperature (Note 15).
b. Cytokine production is evaluated by enzyme-linked immunosorbent assay (ELISA). Examples of ELISA kits include: human IFN gamma, human TNF alpha, human IL-2 and human IL-6 (Thermo Fisher, Waltham, MA). Non pre-coated ELISA kits are recommended to reduce background reading. Refer to the vendors' manuals and instructions for specific procedures of ELISA (Note 16).

9 Concluding remarks

These two assays provide convenient and robust approaches to evaluate CAR T cell function in the setting of high tumor burden as a measure of potency. Utilizing this *in vitro* stress-test, these assays recapitulate the high tumor burden associated with *in vivo* tumor models. Both assays are able to create the environment to identify the acquisition of an exhausted T cell state in which the therapeutic product loses the capability to proliferate and recursively kill despite a potent initial response. These assays also provide an integrated approach to compare multiple effector functions, including tumor cell killing, T cell expansion, cell phenotype and cytokine production. While the ELTK assay reduces manipulation time of evaluating co-culture conditions of high tumor burden, the rechallenge assay allows for the detection of killing, phenotypes and cytokines at specific time periods, which is critical for monitoring T cell dysfunction, including defects in cytokine production prior to the reduction of cytotoxicity (Wherry & Kurachi, 2015).

These assays can be scaled-up and combined with multi-omics analyses of CAR T cells (Salter et al., 2018; Wang et al., 2018), and T cell polyfunctionality, which has been adopted to predict clinical responses (Rossi et al., 2018). Tumor cells can also be tested for their adaptive resistance responses to T cell activity, such as up-regulation of PD-L1 (Garcia-Diaz et al., 2017). Since samples are harvested at multiple time points, the re-challenge assay allows for not only static, but also dynamic analysis of CAR T cell behavior when responding to excess number of tumor cells. These assays are particularly powerful in assessing the effector potency of CAR T cells targeting the same antigen, to compare different CAR designs, manufacturing processes and/or differences between patient-specific products. We have used these assays, which were predictive of differential *in vivo* antitumor function between CD4+ and CD8+ CAR T cells (Fig. 5) (Wang et al., 2018). Further, these assays also provided informative results, which were consistent with *in vivo* antitumor responses, for the assessment of how tumor PD-L1 expression affects CAR T cell potency (Portnow et al., 2020) and for the functional improvement of CAR T cells with additional genetic modifications (Wang et al., 2021).

Clinical responses of CAR T cells have shown substantial inter-patient variation (Fraietta, Lacey, et al., 2018; Fraietta, Nobles, et al., 2018; Rossi et al., 2018). Functional characterization of bespoke CAR T cell products is an essential aspect of these clinical studies. Meanwhile, in addition to the reported phenotypic evaluations on the infusion products (Deng et al., 2020; Fraietta, Lacey, et al., 2018), the post-stimulation phenotypes also inform the fitness of CAR T cells. The assays described here allow for higher-throughput screening of clinical products in comparison to *in vivo* tumor models, while retaining similar predictive fidelity. These assays could be used as functional tests for CAR T cell design and pre-clinical and clinical development.

10 Notes

Note 1. Selection of co-culture media is believed to have minimal impact on the outcomes. We have tested both T cell co-culture media (for Assay #1) and DMEM:F10 based co-culture media (for Assay #2), which resulted in similar outcomes. All co-culture medias are used without addition of exogenous cytokines.

Note 2. Although X-VIVO 15 can be used as a serum-free media, we find the addition of 10% FCS is optimal for ex vivo T cell expansion.

Note 3. Cytokine (i.e., rhIL-2 and rhIL-15) supplemented media, used for T cell thawing and expansion, should be added fresh to the media and replenished every 2–3 days.

Note 4. Resting overnight ensures that viable cells are used in assay.

Note 5. CAR engineering efficiency is a critical parameter and needs to be determined via flow cytometry before proceeding to potency assays. Refer to Sections 6–7 for detailed procedures of flow cytometry on CAR T cells.

Note 6. The selection of target cells is based on the expression of CAR-targeted antigen. Refer to specific procedures of maintaining and dissociation/digestion for different target cell lines. Here, we describe the protocol using our clinically-validated IL13Rα2-targeted CAR T cells and IL13Rα2-expressing brain tumor cells as an example (Brown et al., 2017).

Note 7. It is highly recommended that the viability of both effector and target cells are >70% for these assays.

Note 8. The concentration of tumor cells used for both assays may be subject to change based on different CAR T cells tested and the selected target cells. With the CAR T cells used here, we previously determined that these CAR T cells can eliminate a 1:4 ratio of tumor cells and also double in number during the first 48 h, which led to the setup of the ratios for rechallenge. Before applying the assay to new CAR T cell-tumor combinations, a pilot study to evaluate the killing and expansion of CAR T cells during the first 48 h is recommended. The optimal condition should allow for the most potent CAR T cells tested in the assay to eliminate >80% of tumor cells at the end of the first tumor challenge (48 h).

Note 9. The transfer and discarding of co-culture supernatant should be done under sterile conditions inside a biocontainment hood.

Note 10. It is recommended to include at least 3 time points of analysis (days 1, 3 and 5) each with at least 3 replicating analysis wells.

Note 11. Supernatants can be removed by a quick flick of the plate which we have found does not disturb the cell pellet. Do not use force (e.g., bang the plate) when discarding the supernatant.

Note 12. Both anti-CAR and anti-transgene staining can be used to determine CAR expression. The readout of both methods should be similar when analyzing CAR T cells without tumor cell co-culture. Stimulated CAR T cells, however, may undergo CAR internalization (Walker et al., 2017), therefore, anti-transgene can be more reflective of CAR-transduced cells.

Note 13. PD-1, LAG-3 and TIM-3 are all T cell exhaustion markers, yet the expression of a single marker may also indicate T cell activation. Therefore, we recommend using the co-expression of all three markers to evaluate T cell exhaustion.

Note 14. After target cell stimulation, CAR T cells primarily display central memory and effector memory T cell phenotypes.

Note 15. Samples from the initial setup of the rechallenge assay and total wells plated should cover 2 (days 2 and 4 of a 5-day assay) or 3 (days 2, 4 and 6 of a 7-day assay) timepoints.

Note 16. During the process of tumor cell rechallenge, the ability of CAR T cells to produce cytokines is usually reduced over time. Thus, the samples harvested at later time points will require less dilution to be accurately read in the same assay plate.

References

Akhavan, D., Alizadeh, D., Wang, D., Weist, M. R., Shepphird, J. K., & Brown, C. E. (2019). CAR T cells for brain tumors: Lessons learned and road ahead. *Immunological Reviews, 290*, 60–84.

Brentjens, R. J., Davila, M. L., Riviere, I., Park, J., Wang, X., Cowell, L. G., et al. (2013). CD19-targeted T cells rapidly induce molecular remissions in adults with chemotherapy-refractory acute lymphoblastic leukemia. *Science Translational Medicine, 5*, 177ra138.

Brown, C. E., Aguilar, B., Starr, R., Yang, X., Chang, W. C., Weng, L., et al. (2017). Optimization of IL13Ralpha2-targeted chimeric antigen receptor T cells for improved anti-tumor efficacy against Glioblastoma. *Molecular Therapy: The Journal of the American Society of Gene Therapy, 26*(1), 31–44.

Brunner, K. T., Mauel, J., Cerottini, J. C., & Chapuis, B. (1968). Quantitative assay of the lytic action of immune lymphoid cells on 51-Cr-labelled allogeneic target cells in vitro; inhibition by isoantibody and by drugs. *Immunology, 14*, 181–196.

Cherkassky, L., Morello, A., Villena-Vargas, J., Feng, Y., Dimitrov, D. S., Jones, D. R., et al. (2016). Human CAR T cells with cell-intrinsic PD-1 checkpoint blockade resist tumor-mediated inhibition. *The Journal of Clinical Investigation, 126*, 3130–3144.

Deng, Q., Han, G., Puebla-Osorio, N., Ma, M. C. J., Strati, P., Chasen, B., et al. (2020). Characteristics of anti-CD19 CAR T cell infusion products associated with efficacy and toxicity in patients with large B cell lymphomas. *Nature Medicine, 26*, 1878–1887.

Eyquem, J., Mansilla-Soto, J., Giavridis, T., van der Stegen, S. J., Hamieh, M., Cunanan, K. M., et al. (2017). Targeting a CAR to the TRAC locus with CRISPR/Cas9 enhances tumour rejection. *Nature, 543*, 113–117.

Fesnak, A. D., June, C. H., & Levine, B. L. (2016). Engineered T cells: The promise and challenges of cancer immunotherapy. *Nature Reviews Cancer, 16*, 566–581.

Fraietta, J. A., Lacey, S. F., Orlando, E. J., Pruteanu-Malinici, I., Gohil, M., Lundh, S., et al. (2018). Determinants of response and resistance to CD19 chimeric antigen receptor (CAR) T cell therapy of chronic lymphocytic leukemia. *Nature Medicine, 24*, 563–571.

Fraietta, J. A., Nobles, C. L., Sammons, M. A., Lundh, S., Carty, S. A., Reich, T. J., et al. (2018). Disruption of TET2 promotes the therapeutic efficacy of CD19-targeted T cells. *Nature, 558*, 307–312.

Garcia-Diaz, A., Shin, D. S., Moreno, B. H., Saco, J., Escuin-Ordinas, H., Rodriguez, G. A., et al. (2017). Interferon receptor signaling pathways regulating PD-L1 and PD-L2 expression. *Cell Reports, 19*, 1189–1201.

Gattinoni, L., Klebanoff, C. A., Palmer, D. C., Wrzesinski, C., Kerstann, K., Yu, Z., et al. (2005). Acquisition of full effector function in vitro paradoxically impairs the in vivo antitumor efficacy of adoptively transferred CD8+ T cells. *The Journal of Clinical Investigation, 115*, 1616–1626.

Grupp, S. A., Kalos, M., Barrett, D., Aplenc, R., Porter, D. L., Rheingold, S. R., et al. (2013). Chimeric antigen receptor-modified T cells for acute lymphoid leukemia. *The New England Journal of Medicine, 368*, 1509–1518.

Lee, D. W., Kochenderfer, J. N., Stetler-Stevenson, M., Cui, Y. K., Delbrook, C., Feldman, S. A., et al. (2015). T cells expressing CD19 chimeric antigen receptors for acute lymphoblastic leukaemia in children and young adults: A phase 1 dose-escalation trial. *Lancet, 385*, 517–528.

Long, A. H., Haso, W. M., Shern, J. F., Wanhainen, K. M., Murgai, M., Ingaramo, M., et al. (2015). 4-1BB costimulation ameliorates T cell exhaustion induced by tonic signaling of chimeric antigen receptors. *Nature Medicine, 21*, 581–590.

Malandro, N., Budhu, S., Kuhn, N. F., Liu, C., Murphy, J. T., Cortez, C., et al. (2016). Clonal abundance of tumor-specific CD4(+) T cells potentiates efficacy and alters susceptibility to exhaustion. *Immunity, 44*, 179–193.

Porter, D. L., Hwang, W. T., Frey, N. V., Lacey, S. F., Shaw, P. A., Loren, A. W., et al. (2015). Chimeric antigen receptor T cells persist and induce sustained remissions in relapsed refractory chronic lymphocytic leukemia. *Science Translational Medicine, 7*, 303ra139.

Portnow, J., Wang, D., Blanchard, M. S., Tran, V., Alizadeh, D., Starr, R., et al. (2020). Systemic anti–PD-1 immunotherapy results in PD-1 blockade on T cells in the cerebrospinal fluid. *JAMA Oncology, 6*, 1947–1951.

Priceman, S. J., Forman, S. J., & Brown, C. E. (2015). Smart CARs engineered for cancer immunotherapy. *Current Opinion in Oncology, 27*, 466–474.

Priceman, S. J., Gerdts, E. A., Tilakawardane, D., Kennewick, K. T., Murad, J. P., Park, A. K., et al. (2018). Co-stimulatory signaling determines tumor antigen sensitivity and persistence of CAR T cells targeting PSCA+ metastatic prostate cancer. *Oncoimmunology, 7*, e1380764.

Rossi, J., Paczkowski, P., Shen, Y. W., Morse, K., Flynn, B., Kaiser, A., et al. (2018). Preinfusion polyfunctional anti-CD19 chimeric antigen receptor T cells are associated with clinical outcomes in NHL. *Blood, 132*, 804–814.

Salter, A. I., Ivey, R. G., Kennedy, J. J., Voillet, V., Rajan, A., Alderman, E. J., et al. (2018). Phosphoproteomic analysis of chimeric antigen receptor signaling reveals kinetic and quantitative differences that affect cell function. *Science Signaling, 11*, 1–17.

Vishwanath, R. P., Brown, C. E., Wagner, J. R., Meechoovet, H. B., Naranjo, A., Wright, C. L., et al. (2005). A quantitative high-throughput chemotaxis assay using bioluminescent reporter cells. *Journal of Immunological Methods, 302*, 78–89.

Walker, A. J., Majzner, R. G., Zhang, L., Wanhainen, K., Long, A. H., Nguyen, S. M., et al. (2017). Tumor antigen and receptor densities regulate efficacy of a chimeric antigen receptor targeting anaplastic lymphoma kinase. *Molecular Therapy, 25*, 2189–2201.

Wang, D., Aguilar, B., Starr, R., Alizadeh, D., Brito, A., Sarkissian, A., et al. (2018). Glioblastoma-targeted CD4 + CAR T cells mediate superior antitumor activity. *JCI Insight, 3*(10), 1–18.

Wang, D., Prager, B. C., Gimple, R. C., Aguilar, B., Alizadeh, D., Tang, H., et al. (2021). CRISPR screening of CAR T cells and Cancer stem cells reveals critical dependencies for cell-based therapies. *Cancer Discovery, 11*, 1192–1211.

Wang, D., Starr, R., Alizadeh, D., Yang, X., Forman, S. J., & Brown, C. E. (2019). In vitro tumor cell Rechallenge for predictive evaluation of chimeric antigen receptor T cell antitumor function. *Journal of Visualized Experiments: JoVE, 144*, 1–7.

Wherry, E. J., & Kurachi, M. (2015). Molecular and cellular insights into T cell exhaustion. *Nature Reviews. Immunology, 15*, 486–499.

Printed in the United States
by Baker & Taylor Publisher Services